DNA Arrays

METHODS IN MOLECULAR BIOLOGY™

John M. Walker, Series Editor

METHODS IN MOLECULAR BIOLOGY™

DNA Arrays
Methods and Protocols

Edited by

Jang B. Rampal

Beckman Coulter, Inc., Fullerton, CA

Humana Press ✳ Totowa, New Jersey

© 2001 Humana Press Inc.
999 Riverview Drive, Suite 208
Totowa, New Jersey 07512

This publication is printed on acid-free paper. ⧄
ANSI Z39.48-1984 (American Standards Institute)
Permanence of Paper for Printed Library Materials.

Cover design by Patricia F. Cleary.
Production Editor: Jessica Jannicelli

For additional copies, pricing for bulk purchases, and/or information about other Humana titles, contact Humana at the above address or at any of the following numbers: Tel.: 973-256-1699; Fax: 973-256-8341; E-mail: humana@humanapr.com; or visit our Website: http://humanapress.com

Photocopy Authorization Policy:

Printed in the United States of America. 10 9 8 7 6 5 4 3 2

Library of Congress Cataloging in Publication Data

Main entry under title:

Methods in molecular biology™.

DNA arrays: methods and protocols / edited by Jang B. Rampal.
    p. ; cm.—(Methods in molecular biology ; 170)
  Includes bibliographical references and index.
  ISBN 0-89603-822-X (alk. paper)
  1. DNA microarrays—Laboratory manuals. I. Rampal, Jang B. II. Methods in molecular biology (Totowa, NJ);
  v . 170
  DNLM: 1. XXX. 2. XXX. QW 165.5P2 S969 2001]
  QP624.5D726 D62 2001
  572.8'6—dc21

                                     00-040791

Preface

Microarray technology provides a highly sensitive and precise technique for obtaining information from biological samples, with the added advantage that it can handle a large number of samples simultaneously that may be analyzed rapidly. Researchers are applying microarray technology to understand gene expression, mutation analysis, and the sequencing of genes. Although this technology has been experimental, and thus has been through feasibility studies, it has just recently entered into widespread use for advanced research.

The purpose of *DNA Arrays: Methods and Protocols* is to provide instruction in designing and constructing DNA arrays, as well as hybridizing them with biological samples for analysis. An additional purpose is to provide the reader with a broad description of DNA-based array technology and its potential applications. This volume also covers the history of DNA arrays—from their conception to their ready off-the-shelf availability—for readers who are new to array technology as well as those who are well versed in this field. Stepwise, detailed experimental procedures are described for constructing DNA arrays, including the choice of solid support, attachment methods, and the general conditions for hybridization.

With microarray technology, ordered arrays of oligonucleotides or other DNA sequences are attached or printed to the solid support using automated methods for array synthesis. Probe sequences are selected in such a way that they have the appropriate sequence length, site of mutation, and T_m. The target biological sample is selected for the disease of interest by amplifying that particular sequence by PCR or other techniques. This amplified DNA target is made to hybridize with presynthesized sequences on solid supports. Hybridized arrays are read with CCD cameras and reports are generated with computer-aided technology.

The first chapter by Professor Southern describes a brief history of DNA array technology followed by two more chapters (2, 3) giving detailed reviews of basic principles in specific areas of interest. Chapter 4 deals with ethical issues related to genetic analysis. Chapter 5 describes a unique way of synthesizing arrays using the photolithographic approach; it also includes a discussion of the synthesis of modified monomers and their use. Chapter 6 demonstrates genotyping using DNA Mass Array™ methodology. The next two chapters (7, 8) mainly discuss printing or spotting technologies for array synthesis. Chapters 9 and 10 discuss sample preparation (DNA, RNA) and the conditions used during hybridization. Chapter 11 deals with sequence analysis using sequencing-by-hybridization (SBH). Chapter 12 provides information on antisense reagents, a future drug market that will be used to study the effect of these molecules by using array hybridization. Chapter 13 specifically describes HLA-DQA typing techniques. Application of array technologies in gene expression analysis is highlighted in Chapter 14. These technologies go one step further toward making it possible for the expression of genes via DNA arrays. Chapter 15 is devoted to data extraction and data analysis, also known as bioinformatics. Chapter 16 focuses on application of confocal microscopes in detecting microspots. Chapter 17 discusses commercialization and business aspects of biochip technology.

Once again, we think *DNA Arrays: Methods and Protocols* will provide information to all levels of scientists from novice to those intimately familiar with array technology. We would like to thank all the contributing authors for providing manuscripts. I thank John Walker for editorial guidance and the staff of Humana Press in making it possible to include a large body of available DNA microarray technologies in one single volume. Finally, my thanks to my family, especially to Sushma Rampal who is the light of my life and who is solely responsible for my happiness on this earth, and colleagues for their help in completing this volume.

Jang B. Rampal

Contents

Contributors

BOGDAN ANTOHE • *MicroFab Technologies Inc., Plano, TX*

JOERG BAIER • *Hyseq Inc., Sunnyvale, CA*

KENNETH L. BEATTIE • *Oak Ridge National Laboratory, Oak Ridge, TN*

CHARLES R. CANTOR • *SEQUENOM Inc., San Diego, CA*

ULRICH CERTA • *F. Hoffmann-La Roche Ltd., Roche Genetics, Basel, Switzerland*

GLORIA CHUI • *Hyseq Inc., Sunnyvale, CA*

DAN COLEMAN • *Hyseq Inc., Sunnyvale, CA*

PATRICK COOLEY • *MicroFab Technologies Inc., Plano, TX*

SAVVAS DAMASKINOS • *Biomedical Photometrics Inc., Waterloo, Ontario, Canada*

DAT D. DAO • *DNA Technology Laboratory, Houston Advanced Research Center, The Woodlands, TX*

ANTOINE DE SAIZIEU • *F. Hoffmann-La Roche Ltd., Roche Genetics, Basel, Switzerland*

ROBERT DIAZ • *Hyseq Inc., Sunnyvale, CA*

ARTHUR E. DIXON • *Biomedical Photometrics Inc., Waterloo, Ontario, Canada*

RADOJE DRMANAC • *Hyseq Inc., Sunnyvale, CA*

SNEZANA DRMANAC • *Hyseq Inc., Sunnyvale, CA*

JACQUELINE A. FIDANZA • *Affymetrix Inc., Santa Clara, CA*

DAVID W. GALBRAITH • *Department of Plant Sciences, University of Arizona, Tucson, AZ*

DARRYL GIETZEN • *Hyseq Inc., Sunnyvale, CA*

WAYNE W. GRODY • *Divisions of Molecular Pathology and Medical Genetics, Departments of Pathology and Laboratory Medicine and Pediatrics, UCLA School of Medicine, Los Angeles, CA*

SARAH H. HADDOCK • *DNA Technology Laboratory, Houston Advanced Research Center, The Woodlands, TX*

DEBRA HINSON • *MicroFab Technologies Inc., Plano, TX*

AARON HOU • *Hyseq Inc., Sunnyvale, CA*

HUI JIN • *Hyseq Inc., Sunnyvale, CA*

CHRISTIAN JURINKE • *SEQUENOM GmbH, Hamburg, Germany*

PETER KALOCSAI • *BioDiscovery Inc., Los Angeles, CA*

HUBERT KÖSTER • *SEQUENOM Inc., San Diego, CA*

JIŘÍ MACAS • *Institute of Plant Molecular Biology, České Budějovice, Czech Republic*

ROGELIO MALDONADO-RODRIGUEZ • *Escuela Nacional de Ciencias Biologicas, I.P.N., México*

GLENN H. MCGALL • *Affymetrix Inc., Santa Clara, CA*

ANDREI MIRZABEKOV • *Biochip Technology Center, Argonne National Laboratory, Argonne, IL; and Joint Human Genome Project, Engelhardt Institute of Molecular Biology, Russian Academy of Sciences, Moscow, Russia*

JAN MOUS • *F. Hoffmann-La Roche Ltd., Roche Genetics, Basel, Switzerland*

MARCELA NOUZOVÁ • *Institute of Plant Molecular Biology, České Budějovice, Czech Republic*

CATHERINE O'CONNELL • *Biotechnology Division, National Institute of Standards and Technology, Gaithersburg, MD*

ELIZABETH A. PIERSON • *Department of Plant Sciences, University of Arizona, Tucson, AZ*

CHRISTINE QUARTARARO • *DNA Technology Laboratory, Houston Advanced Research Center, The Woodlands, TX*

JANG B. RAMPAL • *Beckman Coulter, Inc., Fullerton, CA*

HENRY RODRIGUEZ • *Biotechnology Division, National Institute of Standards and Technology, Gaithersburg, MD*

KENNETH E. RUBENSTEIN • *The Lion Consulting Group, Emeryville, CA*

CHRIS SEIDEL • *Operon Technologies Inc., Alameda, CA*

SOHEIL SHAMS • *BioDiscovery Inc., Los Angeles, CA*

RALPH SINIBALDI • *Operon Technologies Inc., Alameda, CA*

MUHAMMAD SOHAIL • *University of Oxford, Department of Biochemistry, South Parks Road, Oxford, UK*

EDWIN M. SOUTHERN • *University of Oxford, Department of Biochemistry, South Parks Road, Oxford, UK*

HANS-JOCHEN TROST • *MicroFab Technologies Inc., Plano, TX*

TATJANA UKRAINCZYK • *Hyseq Inc., Sunnyvale, CA*

DIRK VAN DEN BOOM • *SEQUENOM GmbH, Hamburg, Germany*

DAVID WALLACE • *MicroFab Technologies Inc., Plano, TX*

CHONGJUN XU • *Hyseq Inc., Sunnyvale, CA*

WENYING XU • *Department of Plant Sciences, University of Arizona, Tucson, AZ*

JORDANKA ZLATANOVA • *Biochip Technology Center, Argonne National Laboratory, Argonne, IL*

1

DNA Microarrays

History and Overview

Edwin M. Southern

1. Introduction
1.1. From Double Helix to Dot Blots

It may seem premature to be writing a history of DNA microarrays because this technology is relatively new and clearly has more of a future than a past. However readers could benefit from learning something about the technical basis of DNA microarrays, and younger readers may be curious to know something of the origins and antecedents of this new technology. In this chapter, I have attempted also a critical overview of the current state of the art.

Soon after the first description of the double helix by Watson and Crick *(1)*, it was shown that the two strands could be separated by heat or treatment with alkali. The reverse process, which underlies all the methods based on DNA renaturation or molecular hybridization, was first described by Marmur and Doty *(2)*. It was quickly established that the two sequences involved in duplex formation must have some degree of sequence complementarity, and that the stability of the duplex formed depends on the extent of complementarity. These remarkable properties suggested ways to analyze relationships between nucleic acid sequences, and analytical methods based on molecular hybridization were rapidly developed and applied to a range of biological problems. Some methods, such as those developed by Nygaard and Hall *(3)* and Gillespie and Spiegelman *(4)*, measured the end point or the rate of interaction between an RNA molecule and the DNA from which it was transcribed. This was then used to measure the number of repeated sequences such as ribosomal genes using labeled rRNA as probe and to measure the concentration of RNAs in

From: *Methods in Molecular Biology, vol. 170: DNA Arrays: Methods and Protocols*
Edited by: J. B. Rampal © Humana Press Inc., Totowa, NJ

solution. These were early forerunners of the current application of DNA microarrays to the analysis of sequence diversity and levels of gene expression.

In the late 1960s, Pardue and Gall *(5)* and Jones and Robertson *(6)* discovered a way of locating the position of specific sequences in the nucleus or chromosomes by carrying out the hybridization reaction on cells fixed to microscope slides (*in situ* hybridization, now more familiarly known as fluorescence *in situ* hybridization [FISH], following the introduction of fluorescent probes). The method used to fix chromosomes and nuclei to microscope slides in a way that allowed the DNA to take part in duplex formation with the probe is now used to fix DNA spotted on to slides in one microarray method. And the multicolor fluorescent labeling techniques introduced by Ried et al. *(7)* and Balding and Ward *(8)*, for the analysis of multiple probes by FISH, are now used for comparative analysis of mRNAs from different sources.

In the mid-1970s, recombinant DNA methods were being developed, and although the great potential of the methods was widely recognized, this could not be realized fully without ways of detecting specific sequences in recombinant clones. Grunstein and Hogness *(9)* provided the means to do this by applying molecular hybridization directly to bacterial colonies lysed and fixed to a membrane; later, Benton and Davis *(10)* devised a related method for phage plaques. These methods had a tremendous influence on the rate of discovery of new genes.

1.2. Large-Scale Analysis

Bacteria or yeast cells carrying recombinant DNAs are spread randomly onto plates for cloning. Large sets of clones were picked to be organized and stored as "libraries" in microtiter plates. Some of these libraries became standards that were used repeatedly by researchers looking for specific genes. Eventually, some of the libraries were analyzed to find sets of overlapping clones to create the physical maps that have been so important for positional cloning of genes by reverse genetics and have provided substrates for genome sequencing. In the late 1980s, Hoheisel et al. *(11)* took the organization a stage further and promoted the idea of using multiple libraries arrayed on filters at high density as tools for cross-correlating cloned sequences. The technique of analyzing multiple hybridization targets in parallel by applying them to a filter in a defined pattern, the familiar dot blot, was introduced by Kafatos et al. *(12)*. In this procedure, not only are the hybridizations carried out in parallel, simplifying the process and ensuring reproducibility, but imaging methods allow for parallel measurement of signals as well. Parallel processing through a series of processes is an important feature of all array-based methods. Hoheisel et al. *(11)* increased the density of spots by replacing the manual procedures used to pick and spot clones onto filters by robotics. Automation increased the speed of the operation, removed human errors that inevitably occur in with highly repetitive procedures, and improved the accuracy of placing samples. This was a first step toward microarrays.

1.3. Synthetic DNA

During this period, organic chemistry also underwent a revolution, fueled by the introduction of solid-phase synthesis *(13)*. Its impact was felt in molecular biology, which benefited from the development by Letsinger et al. *(14)* and Beaucage and Caruthers *(15)*, of methods that were suitable for the solid-phase synthesis of nucleic acids. These new methods built on the pioneering work of Khorana et al. *(16)*, who had demonstrated the possibility of synthesizing complex nucleic acids, using methods developed by Corby et al. *(17)* in the 1950s. It is now possible to synthesize, by automated push-button methods, polynucleotides of any sequence up to a limit determined by the coupling yield at each step; DNA molecules in excess of 200 nucleotide residues have been made by these methods. Wallace et al. *(18)* and Conner et al. *(19)* introduced synthetic oligonucleotides as hybridization probes in 1979 and subsequently used them to analyze mutations. The same chemistry provided the primers needed for the polymerase chain reaction (PCR), first proposed by Kleppe et al. *(20)* and reduced to practice by Mullis et al. *(21)*.

2. Dot Blots, Reverse Dot Blots, and Microarrays

What distinguishes a DNA microarray from a dot blot? In the dot-blot format described by Kafatos et al. *(12)*, multiple targets are arrayed on the support (here the term *probe* is used for the nucleic acids of known sequence, which will be attached to the surface in the case of the microarray, and the term *target* describes the unknown sequence or collection of sequences to be analyzed); the probe, normally a single sequence, is labeled and applied under hybridization conditions to the membrane. Saiki et al. *(22)* introduced a variant, the *reverse dot blot*, in which multiple probes are attached as an array to the membrane and the target to be analyzed is labeled. Similar in practice, each method has quite different applications. The first arrays made on impervious supports were made in my laboratory by Maskos *(23)* at about the same time the reverse dot blot was reported. These arrays comprised short oligonucleotides—up to 19-mer—synthesized *in situ (24,25)*. These early experiments established the basis of much of the current array technology and confirmed the important advantages of using impermeable supports.

2.1. Impermeable Supports

Blotting procedures *(26)* necessarily use a porous support, which has some advantages. For example, it is possible to load quite large amounts of nucleic acid on a small area because the pores of the membrane provide a larger total surface for binding. Furthermore, the nucleic acids can be applied in a relatively large volume as it soaks into the pores of the membrane without excessive lateral spreading. However, the boundaries and shapes of the spots are poorly defined and the amount of oligonucleotide deposited is difficult to control accurately. The demands of genome projects brought the need for analysis on a new, much larger scale, and although it was possible to increase the area of dot blots, it was not possible to reduce the size of spots beyond certain limits, or to control their size and shape on a porous membrane. These factors become crucial for automated analysis of hybridization signals, when it is necessary to locate accurately the positions of the spots and to know in advance their precise shape and size, and an additional, major advantage of glass or plastic supports is their dimensional stability and rigidity. Permeable membranes swell in solvent and tend to shrink and distort when dried; their fragility and flexibility make it difficult to register their position during spotting and reading. Thus, it is not possible to locate spots with the high precision that can be achieved on a rigid substrate.

The introduction of impermeable supports was a major departure that afforded several advantages. As the nucleic acids form a monolayer, saturating the surface, the amount attached is consistent from one region of the array to another, and, as they are on the surface, the nucleic acids are favorably placed to take part in hybridization reactions. Interactions with the solution phase are much faster, because molecules do not have to diffuse into and out of the pores. All stages of the process benefit from this easy access. The target polynucleotides can find immediate access to the probes, accelerating hybridization, and ensuring that the multiple interactions involved in duplex formation are not perturbed by the diffusion process or any steric inhibition that may result from confinement in the pores of a membrane. Washing is also unimpeded by the need for excess labeled material to be diffused out of the pores of a membrane, which speeds up the procedure, improves reproducibility, and reduces background. All these factors are important when the objective is to achieve reliable hybridization signals to the high level of accuracy needed to distinguish small differences in signal from different probes on the array.

Several materials are likely to be suitable as substrates for making arrays. Glass is the material of choice: it is cheap, has good physical characteristics, and is easily modified for covalent attachment or for *in situ* synthesis of nucleic acids. Polypropylene has also been used (27) and has the advantage over glass for some applications in that it is flexible and relatively soft, so that it can be bent to shape, and reaction cells can be sealed against the surface by pressure for one of the modes of *in situ* synthesis. My laboratory and others have used silicon for research applications, but it is an expensive material to use for production. We have found that the nature of the support, and especially the nature of the linkage between the support and the oligonucleotides, greatly affects performance. In particular, we have found that an optimal density and length of linker increases the hybridization yield substantially *(28)*.

Arrays made by deposition or by *in situ* synthesis occasionally perform poorly: the background may be dirty or the hybridization weak or patchy. Experience has shown that poor derivatization of the substrate, prior to attachment or coupling, is one of the main causes of poor performance of an array. The difficulty we are faced with is how to monitor the quality of the product at various stages of manufacturing and to use it in a nondestructive way. The amount of material deposited on the surface of the substrate is a molecular monolayer at most, equivalent to about 10 pmol/mm^2. This is enough material to analyze by sensitive techniques, such as mass spectrometry, capillary electrophoresis, or high-performance liquid chromatogrphy (HPLC). However, the material is covalently bound to the surface, and these methods are not suitable for the analysis of the linker materials. Nondestructive optical methods—ellipsometry and interferometry—have been used successfully to analyze glass surfaces after derivatization with a linker and subsequent oligonucleotide synthesis *(29)*, but these methods are not available to most laboratories. If a cleavable linker is used, the nucleic acid molecule can be analyzed after cleaving it from the support. This method has been used to show the length distribution, and hence estimate step yields, of nucleic acids synthesized *in situ*.

3. Fabrication

3.1. Arrays of Presynthesized Probes

The route to making arrays by spotting probes of cloned sequences, or nucleic acid synthesized by PCR, has been straightforward. The support used for this purpose is the same as that used for *in situ* hybridization: glass slides subbed with poly-L-lysine, to which the probes are covalently crosslinked by ultraviolet irradiation (e.g., for protocols, *see* http://cmgm.stanford.edu/pbrown/). The method of application is an adaptation of a computer-controlled xyz stage with a head carrying a pin or pen device to pick up small drops of solution from the multiwell plates and carry them to the surface. The pens used in these devices are adapted from designs used in ink pens, either metal capillaries or quills. For chemically synthesized nucleic acids, end attachment is favored, and various methods for attachment to solid supports have been used (e.g., *see* **ref. *30***). Quality control is becoming important, especially as nucleic acid arrays enter clinical diagnostic applications, and it is an advantage of presynthesized nucleic acid probes that their quality can be checked before they are attached to the surface.

3.2. In Situ *Synthesis of Probes*

A further benefit of using impermeable supports is that it permits array fabrication by *in situ* synthesis of nucleic acids on the surface. *In situ* synthesis has a number of advantages over deposition of presynthesized probes. It combines the advantages of solid-phase synthesis (high coupling yields and high purity, no need for purification) with those of combinatorial chemistry (a large diversity of compounds can be made in few steps) *(31)*. Typically, the number of coupling steps is a small multiple of the length of probes made on the array. For example, there are combinatorial methods for making all 4^8 octanucleotides that require only eight coupling steps *(32)*. This is to be compared with $8 \times 65,536 = 524,288$ steps if the probes are made individually. Two types of approach were developed to confine the synthesis to small, defined regions of the solid support.

The simpler approach adapted existing chemistry, delivering reagents to confined areas: e.g., using drop-on-demand ink-jet technology *(33)* or irrigating the surface through flow channels *(25,32)*. A more specialized method adapted the photolithographic methods used in the semiconductor industry *(34)* and required the development of new photolabile protecting groups for nucleotide precursors.

3.2.1. Ink-Jet Fabrication

Ink-jet printers, although designed to fire droplets of ink at paper, are readily adapted to firing solutions of nucleotide reagents at a glass surface *(33)*. The main change has been replacement of acetonitrile, the solvent commonly used for oligonucleotide reagents, by a more viscous and less volatile solvent such as adiponitrile. Very small volumes of reagent are delivered at each step. A great advantage of this platform is that the device has much in common with an ink-jet printer, and therefore most of the engineering work had already been done in the development of the printer. As in the printer, pens and the substrate are mounted on drives, which allow accurate relative movement in two axes. The processes of moving the pens and substrate and firing the pens are controlled by a computer using driver software that is easily adapted from printing four colors to delivering precursors for four different bases. For printing, the required sequences are fed to the synthesizer as a text file and converted to instructions to the reagent delivery system. Thus, any set of oligonucleotides can be made by this method, and known sequences can be placed at any position in the array. Reprogramming the system to make a different array is simply a matter of changing the sequence file. The oxidation and deprotection steps and the washes are common to each cycle and are carried out by flooding the whole surface with an excess of reagent or solvent. Thus, the method is flexible and makes economical use of the most expensive reagents.

As would be expected from the high resolution that can be achieved by ink-jet printers, the dimensions of arrays made in this way are small, with cells about 100–150 μ in diameter, at 100–200-μ centers.

3.2.2. Flow Channels and Cells

An alternative way of synthesizing oligonucleotides *in situ* is to confine the reagents to regions defined by pressing open-faced flow channels *(25,32)* or cells against the surface *(35)*. This method is particularly well suited to making arrays of two types: those comprising all oligonucleotides of a given length, and those comprising all the complements of a target of known sequence.

The following protocol illustrates how combinatorial methods can be used to create arrays of all sequences in an economical manner. 4^s oligonucleotides of length s are synthesized in s steps. Linear flow channels are assumed in the protocol, but other shapes can be used, and the order of coupling is not critical. The precursors for the four bases, A, C, G, T, are introduced through channels to make 4 broad stripes of the mononucleotides on a square plate. A second set is laid down in four narrower stripes within each of the monomers to create 16 stripes of dinucleotides. This process is iterated, each time using stripes one quarter the width of the previous set, until the oligonucleotides have reached half their final length. At this point, the plate is turned 90° and the whole process is repeated. The result is an array in which all sequences of the chosen length are represented just once in known positions. The dimensions of such arrays are determined by the width of the stripes. This protocol will generate cells with sides equal to the narrowest channel width. It is possible by micromachining to make flow channels <100 μ wide.

Scanning arrays, comprising a fully overlapping set of oligonucleotides complementary to a target of known sequence, can also be made by economical combinatorial methods *(35)*. In this case, a sealed cell delivers reagents over a circular or diamond-shaped area of the substrate. The cell is displaced along the surface after each coupling by an offset that is a defined fraction, $1/s_{max}$, of the diameter of the circle or the diagonal of the diamond. The bases are coupled in the order in which they occur in the complement of the target sequence. The result is an array that includes all complementary oligonucleotides of length s and also all shorter complements, down to mononucleotides, in the order in which they occur in the target. The size of features is equal to the linear displacement between couplings, which can be small: my laboroatory has made arrays with features <10 μ square using a relatively simple apparatus. Combinatorial synthesis produces arrays with interesting properties. Their layout is particularly favorable for detailed comparison of hybridization behavior, because adjacent oligonucleotides are related in sequence by a single base difference. In the case of the exhaustive arrays made by the aforementioned protocol, each oligonucleotide is surrounded by others in which one of the terminal bases is replaced by another. In the scanning arrays, each oligonucleotide is adjacent to others that differ in length or sequence by loss, addition, or replacement of one terminal base. Subtle differences in hybridization yield are easily discernible when they are side by side.

3.2.3. Light-Directed Fabrication

Photocleavable protecting groups have several uses in organic synthesis (reviewed in **ref.** *36*) and were used by Fodor et al. *(34)* to develop a way of directing the synthesis of oligonucleotides to specific positions on a glass surface by irradiating the surface through a set of patterned photolithographic masks. Each base addition requires a separate mask, so the set for an array of 20-mers would be $4 \times 20 = 80$ in number.

At each step, the surface is irradiated to remove the protecting group on the 5' hydroxyl group of the nucleotide previously added. The surface is then flooded with the coupling agent for the base and the process continued for the next base. Like ink-jet printing, this method has the advantage that it is "random access"; any sequence can be synthesized at any position. A further advantage is the small size of the arrays. Arrays with 65,536 oligonucleotides in an area 1.28×1.28 cm are commercially available. The smaller the size of the array, the smaller the volume needed for hybridization. A disadvantage of the method is that coupling yields (about 95%) *(37)* are lower than for conventional chemicals (>99%). Thus, the yield of a 20-mer will be about 36% as compared with >80%.

4. Processing
4.1. Targets and Labeling

The target nucleic acid to be analyzed can be RNA or DNA, which should preferably be labeled so that the hybrids can be directly detected. PCR, which is commonly used, produces targets that are double stranded and unsuitable for hybridization to oligonucleotides. Asymmetric amplification makes enough single strands, but a better method is to destroy one strand by treatment with exonuclease *(38,39)*. Modifications to one of the PCR primers prevent access of the exonuclease to the strand that it primes. We have found this method to be easy, reliable, and able to produce targets that hybridize well. Alternatively, if an appropriate promoter is incorporated into the sequence of one of the PCR primers, a single-stranded transcript can be made readily by a bacterial polymerase, such as the T7 polymerase *(25)*. This method has several advantages: there is substantial additional amplification as a result of the transcription, and the RNA can be labeled to a high specific activity by incorporating labeled precursors. However, RNA molecules fold as a result of intramolecular base pairing to form stable structures that interfere with the hybridization process—the corresponding structures in DNA are less stable. The problem with RNA can be partly relieved by degrading the transcripts to fragments of a size comparable with that of the oligonucleotide probes. The problem is less severe for arrays of spotted cDNAs because hybridization can be carried out at higher temperatures, which melt the intramolecular base pairing.

Radioactivity is convenient and provides sensitive detection, but it has a wide "shine." This is not a problem with membranes, because the dimensions of the features are such that the image degradation is not significant. However, the degradation is large compared with the features that can be achieved on a smooth glass or plastic surface. Fortunately, these materials are suitable for use with fluorescent labels, and this has become the preferred method of labeling in many laboratories.

4.2. Detection and Quantitation

Radioactive detection has many advantages. It has a wide dynamic range, even with a single exposure, but the range can be extended by varying the exposure time. Quantitation can be very precise. It is easy to label targets to a high specific activity by a number of well-established methods. ^{32}P has a wide shine, but ^{33}P can be imaged by phosphorimaging to a resolution of about 200 μ; in my experience, resolution is limited by the grain structure of the phosphorimager screen. This is satisfactory for cell dimensions of about 1 mm.

Fluorescent labels have different advantages. In particular, they enable double labeling and high-resolution imaging. Confocal microscopy reduces noise by removing out of focus background, but the field of view is limited, and several readers that apply the confocal principle to a large format have been developed for use with arrays and are now on the market.

4.3. Hybridization

The rigid or stiff materials used for microarrays are easier to handle than the membranes used for blotting. In my laboratory, with glass arrays, we find it convenient to place the face of the array against another glass plate and run the hybridisation solution into the gap by capillary action. Alternatively, hybridization can be carried out in a simple cell holding a small volume of liquid. The process is easily automated by housing the array in a flow cell. Precise temperature control is needed for reproducible results, and we have found that the hybridization rate is increased if the hybridization solution is in motion over the surface of the array by, e.g., placing the array in a rotating cylinder.

5. Applications

5.1. Analysis of Sequence Variation: Short Probes

Several areas of biology have benefited greatly from the introduction of methods for analyzing sequence differences. Mapping the human genome using DNA polymorphisms first suggested by Solomon and Bodmer *(40)* and Botstein et al. *(41)* has opened the way for the isolation of a number of disease-causing genes and was a necessary first step toward the present sequencing endeavor. Geneticists studying humans lacked the phenotypic markers that were available to those working with model organisms. Once mapped, large-scale efforts were needed to find the mutations in the candidate genes responsible for the disease phenotype *(42,43)*. DNA polymorphisms, analyzed on a large scale, are expected to give enough analytical power to carry out genetic studies to find the genes associated with common diseases and inherited disease susceptibilities (e.g., *see* **ref. 44**).

Sequence variation is best analyzed with the shortest oligonucleotides that will give specific hybridization to the target site. Lengths much shorter than 15-mer may find cross-reassociations with other sites. On the other hand, it is desirable to use short oligonucleotides for this purpose, to achieve good discrimination between the variants, which, by definition, will be closely related in sequence. This may be difficult with probes much longer than 15-mer. In this length region, it is necessary to carry out hybridization under nonstringent conditions of relatively high salt and low temperature. A problem that can arise is that these conditions also favor intramolecular base pairing in the target, which can prevent hybridization to the short probes *(45)*. This problem can be avoided, to some extent, by using short DNA targets. Another way is to use enzymes, such as polymerase or ligase, in combination with arrays of oligonucleotides.

5.1.2. Enzymes and Chips

The combination of enzymes and chips can be especially useful for the analysis of sequence variation, in which enzymes enhance discrimination beyond what can be achieved by hybridization alone. Polymerases require a primer and incorporate bases one at a time only if they match the complement in the template; the terminal base of the primer must also match that of the template. There are several ways in which the reaction can be used to identify the sequence or a single base at a selected site in the template strand *(46,47)*. Ligases have similar requirements: two oligonucleotides can be joined enzymatically provided they both are complementary to the template at the position of joining *(48)*.

In solid-phase minisequencing, a tethered oligonucleotide is used to capture the target sequence at a position next to a variable base; DNA polymerase and a labeled triphosphate are added and the solution is removed. The identity of the base is determined from the base incorporated *(39,49)*. If fluorescence is used to tag the nucleotide precursors, this method can readily be adapted to multicolor detection. It is an advantage of the enzyme-based methods that the label is incorporated during the test, eliminating the need to label the target.

For analysis of sequence variation at multiple dispersed loci, amplifying all the loci to provide the necessary targets is a most difficult problem.

5.2. Expression Analysis

In contrast to the analysis of a single nucleotide polymorphism, gene expression levels are best analyzed with relatively long probes; most target sequences are likely to be very different in sequence, and, thus, cross-reassociation using long probes will not be a problem. With long probes, it is possible to achieve good yields under stringent hybridization conditions. Hence, it is possible to use a single spot of a PCR product or clone to measure expression levels *(50,51)*, whereas it has proved necessary to use sets of twenty 20-mers for each target to be sure that some would achieve levels of hybridization that are high enough *(52)*.

6. Availability

It has been clear for more than a decade that array-based methods are a key platform for genomics. Few other methods offer their massively parallel scale of analysis. Why has it taken so long for them to be widely adopted? The main reason is that making arrays, although relatively trivial for laboratories with engineering shops, is not easily done by the average biology laboratory. Companies have been slow to enter the market to produce arrays commercially. However, this is changing. This book offers protocols that biologists can use to build their own systems. Several companies are poised to enter the field and make this powerful technology available to the large and growing number of scientists who wish to use it in their endeavors to unlock the huge potential of the emerging genetic resources.

References

1. Watson, J. D. and Crick, F. H. (1953) Molecular structure of nucleic acids: a structure for deoxyribose nucleic acid. G. D. *Nature* **248,** 737–738.
2. Marmur, J. and Doty, P. (1961) Thermal renaturation of deoxyribonucleic acids. *J. Mol. Biol.* **3,** 585–594.
3. Nygaard, A. P. and Hall, B. D. (1964) Formation and properties of RNA-DNA complexes. *J. Mol. Biol.* **9,** 125–142.
4. Gillespie, D. and Spiegelman, S. (1965) A quantitative assay for DNA-RNA hybrids with DNA immobilized on a membrane. *J. Mol. Biol.* **12,** 829–842.

5. Pardue, M. L. and Gall, J. G. (1969) Molecular hybridization of radioactive DNA to the DNA of cytological preparations. *Proc. Natl. Acad. Sci. USA* **64,** 600–604.

6. Jones, K. W. and Robertson, F. W. (1970) Localisation of reiterated nucleotide sequences in Drosophila and mouse by in situ hybridisation of complementary RNA. *Chromosoma* **31,** 331–345.

7. Ried, T., Landes, G., Dackowski, W., Klinger, K., and Ward, D. C. (1992) Multicolor fluorescence in situ hybridization for the simultaneous detection of probe sets for chromosomes 13, 18, 21, X and Y in uncultured amniotic fluid cells. *Human Mol. Genet.* **1,** 307–313.

8. Baldini, A. and Ward, D. C. (1991) In situ hybridization banding of human chromosomes with Alu-PCR products: a simultaneous karyotype for gene mapping studies. *Genomics* **9,** 770–774.

9. Grunstein, M. and Hogness, D. S. (1975) Colony hybridization: a method for the isolation of cloned DNAs that contain a specific gene. *Proc. Natl. Acad. Sci. USA* **72,** 3961–3965.

10. Benton, W. D. and Davis, R. W. (1977) Screening lambdagt recombinant clones by hybridization to single plaques in situ. *Science* **196,** 180–182.

11. Hoheisel, J. D., Ross, M. T., Zehetner, G., and Lehrach, H. (1994) Relational genome analysis using reference libraries and hybridisation fingerprinting. *J. Biotechnol.* **35,** 121–134.

12. Kafatos, F. C., Jones, C. W., and Efstratiadis, A. (1979) Determination of nucleic acid sequence homologies and relative concentrations by a dot hybridization procedure. *Nucleic Acids Res.* **7,** 1541–1552.

13. Merrifield, R. B. (1969) Solid-phase peptide synthesis. *Adv. Enzymol. Relat. Areas Mol. Biol.* **32,** 221–296.

14. Letsinger, R. L., Finnan, J. L., Heavner, W. B., and Lunsford, W. B. (1975) Phosphite coupling procedure for generating internucleotide links. *J. Amer. Chem. Soc.* **97,** 3278–3279.

15. Beaucage, S. L. and Caruthers, M. H. (1981) Deoxynucleoside phosphoramidites - a new class of key intermediates for deoxypolynucleotide synthesis. *Tetrahedron Letters* **22,** 1859–1862.

16. Khorana, H. G., Buchi, H., Ghosh, H., Gupta, N., Jacob, T. M., Kossel, H., Morgan, R., Narang, S. A., Ohtsuka, E., and Wells, R. D. (1966) Polynucleotide synthesis and the genetic code. *Cold Spring Harb. Symp. Quant. Biol.* **31,** 39–49.

17. Corby, N. S., Kenner, G. W., and Todd, A. R. (1952) Nucleotides. Part XVI. Ribonucleoside-5' phosphites. A new method for the preparation of mixed secondary phosphites. *J. Chem. Soc.* 3669–3675.

18. Wallace, R. B., Shaffer, J., Murphy, R. F., Bonner, J., Hirose, T., and Itakura, K. (1979) Hybridization of synthetic oligodeoxyribonucleotides to phi chi 174 DNA: the effect of single base pair mismatch. *Nucleic Acids Res.* **6,** 3543–3557.

19. Conner, B. J., Reyes, A. A., Morin, C., Itakura, K., Teplitz, R. L., and Wallace, R. B. (1983) Detection of sickle cell beta S-globin allele by hybridization with synthetic oligonucleotides. *Proc. Natl. Acad. Sci. USA* **80,** 278–282.

20. Kleppe, K., Ohtsuka, E., Kleppe, R., Molineux, I., and Khorana, H. G. (1971) Studies on polynucleotides. XCVI. Repair replications of short synthetic DNA's as catalyzed by DNA polymerases. *J. Mol. Biol.* **56,** 341–361.

21. Mullis, K., Faloona, F., Scharf, S., Saiki, R., Horn, G., and Erlich, H. (1986) Specific enzymatic amplification of DNA in vitro: the polymerase chain reaction. *Cold Spring Harb. Symp. Quant. Bio.l* **51,** 263–273.

22. Saiki, R. K., Walsh, P. S., Levenson, C. H., and Erlich, H. A. (1989) Genetic analysis of amplified DNA with immobilized sequence-specific oligonucleotide probes. *Proc. Natl. Acad. Sci. USA* **86,** 6230–6234.

23. Maskos, U. (1991) A novel method of nucleic acid sequence analysis. D. Phil. Thesis, Department of Biochemistry, Oxford University, Oxford, UK, 160.

24. Maskos, U. and Southern, E. M. (1993) A novel method for the analysis of multiple sequence variants by hybridisation to oligonucleotides. *Nucleic Acids Res.* **21,** 2267–2268.

25. Maskos, U. and Southern, E. M. (1993) A novel method for the parallel analysis of multiple mutations in multiple samples. *Nucleic Acids Res.* **21,** 2269–2270.

26. Southern, E. M. (1975) Detection of specific sequences among DNA fragments separated by gel electrophoresis. *J. Mol. Biol.* **98,** 503–517.

27. Matson, R. S., Rampal, J. B., Coassin, P. J. (1994) Biopolymer synthesis on polypropylene supports. I. Oligonucleotides *Anal. Biochem.* **217,** 306–310 (erratum appears in *Anal. Biochem.* 1994; **220(1),** 225).

28. Shchepinov, M. S., Case-Green, S. C., and Southern, E. M. (1997) Steric factors influencing hybridisation of nucleic acids to oligonucleotide arrays. *Nucleic Acids Res.* **25,** 1155–1161.

29. Gray, D. E., Case-Green, S. C., Fell, T. S., Dobson, P. J., and Southern, E. M. (1997) Ellipsometric and interferometric characterisation of DNA probes immobilized on a combinatorial array. *Langmuir* **13,** 2833–2842.

30. Guo, Z., Guilfoyle, R. A., Thiel, A. J., Wang, R., and Smith, L. M. (1994) Direct fluorescence analysis of genetic polymorphisms by hybridization with oligonucleotide arrays on glass supports. *Nucleic Acids Res.* **22,** 5456–5465.

31. Maskos, U. and Southern, E. M. (1992) Parallel analysis of oligodeoxyribonucleotide (oligonucleotide) interactions. I. Analysis of factors influencing oligonucleotide duplex formation. *Nucleic Acids Res.* **20,** 1675–1678.

32. Southern, E. M., Maskos, U., and Elder, J. K. (1992) Analyzing and comparing nucleic acid sequences by hybridization to arrays of oligonucleotides: evaluation using experimental models. *Genomics* **13,** 1008–1017.

33. Blanchard, A. P., Kaiser, R. J., and Hood, L. E. (1996) High density oligonucleotide arrays. *Biosensors and Bioelectronics* **11,** 687–690.

34. Fodor, S. P., Read, J. L., Pirrung, M. C., Stryer, L., Lu, A. T., and Solas, D. (1991) Light-directed, spatially addressable parallel chemical synthesis. *Science* **251,** 767–773.

35. Southern, E. M., Case-Green, S. C., Elder, J. K., Johnson, M., Mir, K.U., Wang, L., and Williams, J. C. (1994) Arrays of complementary oligonucleotides for analysing the hybridisation behaviour of nucleic acids. *Nucleic Acids Res.* **22,** 1368–1373.

36. Pillai, V. N. R. (1980) Photoremovable protecting groups in organic synthesis. *Synthesis* 1–26.

37. Pirrung, M. C., Fallon, L., and McGall, G. (1998) Proofing of photolithographic DNA syntheisis with 3',5'-dimethoxybenzoinyloxycarbonyl-protected deoxynucleoside phosphoramidites. *J. Org. Chem.* **63**, 241–246.

38. Shchepinov, M. S., Udalova, I. A., Bridgman, A. J., and Southern, E. M. (1997) Oligonucleotide dendrimers: synthesis and use as polylabelled DNA probes. *Nucleic Acids Res.* **25**, 4447–4454.

39. Nikiforov, T. T., Rendle, R. B., Goelet, P., Rogers, Y. H., Kotewicz, M. L., Anderson, S., Trainor, G. L., and Knapp, M. R. (1994) Genetic Bit Analysis: a solid phase method for typing single nucleotide polymorphisms. *Nucleic Acids Res.* **22**, 4167–4175.

40. Solomon, E., and Bodmer, W. F. (1979) Evolution of sickle variant gene (letter). *Lancet* **1**, 923.

41. Botstein, D., White, R. L., Skolnick, M., and Davis, R. W. (1980) Construction of a genetic linkage map in man using restriction fragment length polymorphisms. *Am. J. Human Genet.* **32**, 314–331.

42. Kerem, B., Rommens, J. M., Buchanan, J. A., Markiewicz, D., Cox, T. K., Chakravarti, A., Buchwald, M., and Tsui, L. C. (1989) Identification of the cystic fibrosis gene: genetic analysis. *Science* **245**, 1073–1080.

43. Tsui, L. C. (1992) Mutations and sequence variations detected in the cystic fibrosis transmembrane conductance regulator (CFTR) gene: a report from the Cystic Fibrosis Genetic Analysis Consortium. *Human Mutat.* **1**, 197–203.

44. Cargill, M., Altshuler, D., Ireland, J., Sklar, P., Ardlie, K., Patil, N., et al. (1999) Characterization of single-nucleotide polymorphisms in coding regions of human genes. *Nat. Genet.* **22**, 231–238.

45. Mir, K. U. and Southern, E. M. (1999) Determining the influence of structure on hybridization using oligonucleotide arrays (In Process Citation). *Nat. Biotechnol.* **17**, 788–792.

46. Cotton, R. G. (1993) Current methods of mutation detection. *Mutat. Res.* **285**, 125–144.

47. Syvanen, A. C., and Landegren, U. (1994) Detection of point mutations by solid-phase methods. *Hum. Mutat.* **3**, 172–179.

48. Nickerson, D. A., Kaiser, R., Lappin, S., Stewart, J., Hood, L., and Landegren, U. (1990) Automated DNA diagnostics using an ELISA-based oligonucleotide ligation assay. *Proc. Natl. Acad. Sci. USA* **87**, 8923–8927.

49. Pastinen, T., Kurg, A., Metspalu, A., Peltonen, L., and Syvanen, A. C. (1997) Minisequencing: a specific tool for DNA analysis and diagnostics on oligonucleotide arrays. *Genome Res.* **7**, 606–614.

50. Schena, M., Shalon, D., Davis, R.W., and Brown, P. O. (1995) Quantitative monitoring of gene expression patterns with a complementary DNA microarray. *Science* **270**, 467–470 (comments).

51. Shalon, D., Smith, S. J., and Brown, P. O. (1996) A DNA microarray system for analyzing complex DNA samples using two-color fluorescent probe hybridization. *Genome Res.* **6,** 639–645.
52. Lockhart, D. J., Dong, H., Byrne, M. C., Follettie, M. T., Gallo, M. V., Chee, M. S., Mittmann, M., Wang, C., Kobayashi, M., Horton, H., and Brown, E. L. (1996) Expression monitoring by hybridization to high-density oligonucleotide arrays. *Nat. Biotechnol.* **14,** 1675–1680 (comments).

2

Gel-Immobilized Microarrays of Nucleic Acids and Proteins

Production and Application for Macromolecular Research

Jordanka Zlatanova and Andrei Mirzabekov

1. Introduction

Biochips are small platforms with spatially arrayed macromolecules (or pieces thereof) that allow the collection and analysis of large amounts of biological information. The principle of the technology is based on specific molecular recognition interactions between the arrayed macromolecules and the test molecule of interest. Classical examples of such recognition reactions are the interactions between the two complementary strands of a double-helical DNA molecule, between a single-stranded DNA stretch and the messenger RNA copied from it during transcription, between an antigen and an antibody, and between small ligands and their nucleic acid or protein partners.

It has become customary to compare biological microchips with electronic microchips with respect to their ability to perform multiple simple reactions in parallel in a high-throughput fashion. Biochips are expected to revolutionize biology in the same way as the electronic chips revolutionized electronics earlier in the twentieth century. Testimony to such a revolutionary role can be found in recent science polls, which ranked the biochip technology among the 10 most important scientific developments in 1998 *(1)*. There are many different types of biochips *(2)*. This chapter focuses on the biochip developed at the Engelhardt Institute of Molecular Biology in Moscow and the Biochip Technology Center at Argonne National Laboratory, Argonne, IL.

From: *Methods in Molecular Biology, vol. 170: DNA Arrays: Methods and Protocols*
Edited by: J. B. Rampal © Humana Press Inc., Totowa, NJ

2. General Description of the MAGIChip™ Technology

MAGIChips™ (Micro Arrays of Gel-Immobilized Compounds on a Chip) are arrays that we have been developing for the past several years *(3–5)*. This array is based on a glass surface that has small polyacrylamide gel elements affixed to it (**Fig. 1**). The size of the pads can differ from $10 \times 10 \times 5$ µm to $100 \times 100 \times 20$ µm, with volumes ranging from picoliters to nanoliters. Each individual gel element can function as an individual test tube because it is surrounded by a hydrophobic glass surface that prevents exchange of solution among the elements. This property is crucial to performing pad-specific reactions, e.g., polymerase chain reaction (PCR) amplification of the hybridization signal of specific sequences of interest.

The production of such microchips involves the following consecutive steps: creation of the microarray of gel elements (pads) on the glass surface (micromatrix), and application and chemical immobilization of different compounds (probes) onto the gel pads (**Fig. 1**). Once the blank micromatrix has been converted into a microchip containing the immobilized probes, the test sample is added and the reaction of molecular recognition takes place under specified conditions. To be able to monitor the results of such molecular interactions, the test sample needs to be labeled, usually by attaching various kinds of fluorescent labels to it.

Finally, the results of the molecular recognition reaction need to be monitored and analyzed. The type of monitoring instrumentation used depends on the required level of performance, and the type of label attached to the test molecule. The analysis of the reaction patterns is automated using specially designed software. In the next sections, we describe in more detail the separate steps of the production and use of the biochip. We also describe some specific features of the different types of biochips—oligonucleotide, cDNA, and protein chips—giving specific examples of their application. Because our efforts have been focused so far on nucleic acid biochips, most of what follows applies to those chips. Some developments concerning protein biochips are described at the end of this chapter.

2.1. Production of the Micromatrix

The matrix of glass-attached gel elements is prepared by photopolymerization *(6)*. The acrylamide solution to be polymerized is applied to a manually assembled polymerization chamber consisting of a quartz mask, two Teflon spacers, and a microscopic glass slide, clamped together by two metal clamps (**Fig. 2A**). The internal side of the quartz mask has ultraviolet (UV)-transparent windows arranged in a specified spatial manner in nontransparent 1-µm-thick chromium film (**Fig. 2B**). The assembled chamber containing the

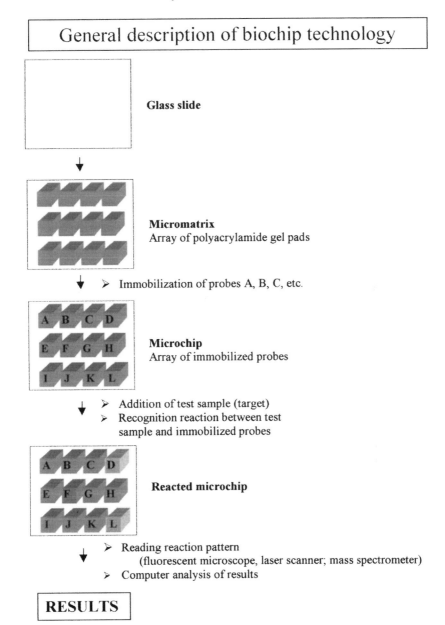

Fig. 1. Overall scheme of the MAGIChip™ technology.

acrylamide solution is exposed to UV light to allow polymerization in only those positions of the chamber that are situated directly under the transparent

A

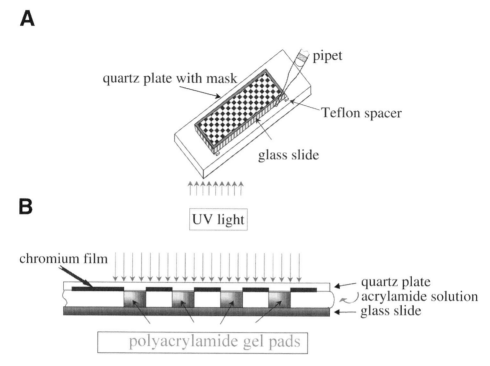

Fig. 2. (**A**) Scheme of the polymerization chamber. (**B**) Scheme of the photopoly-merization.

windows. Following polymerization, the chamber is disassembled, and the matrix is washed, dried, and kept at room temperature in sealed chambers.

We have recently introduced a method for production of matrices that combines the polymerization step with the step of probe immobilization (*7*). In this method acrylamide is copolymerized with oligonucleotides or proteins containing unsaturated residues. In the case of oligonucleotides, such unsaturated units are incorporated during standard phosphoramidite synthesis; in the case of proteins, the protein is chemically attached to the acrylamide monomer containing double bond.

2.2. Probe Activation

Oligonucleotides or DNA fragments to be immobilized in the gel elements should be activated to contain chemically reactive groups for coupling with the activated gel elements. The chemistry of probe activation is chosen in concert with the chemistry of activation of the polyacrylamide gels. Thus, e.g., immo-

A   **B**

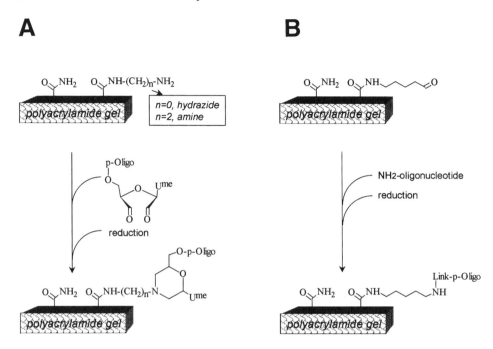

Fig. 3. Chemistry of immobilization of oligonucleotide probes into polyacrylamide gel pads.

bilization in aldehyde-containing gels would require the probe to be functionalized by the introduction of amino groups *(8)* (**Fig. 3**). If the gels are activated by the introduction of amino groups, the probes may be oxidized to contain free aldehyde groups *(9)* (**Fig. 3**). The probe can be prepared by introduction of chemically active groups in terminal positions of the oligonucleotides during their chemical synthesis; alternatively, active groups can be introduced within the chain of nucleotides (chemically synthesized or naturally occurring) in a number of ways *(8,10)*. The probe activation chemistry is well developed and allows for high-yield, reproducible coupling with the gel matrix.

2.3. Application of Probes to Micromatrix and Their Chemical Immobilization in Gel Pads

Routinely, the probes for immobilization are transferred into the gel elements of the micromatrix using a home-designed dispensing robot *(11)*. The fiber-optic pin of the robot has a hydrophobic side surface and a hydrophilic tip surface, and operates at dew point temperature to prevent evaporation of sample

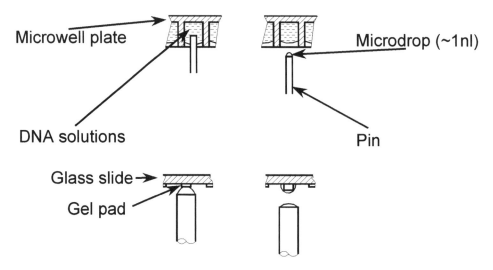

Fig. 4. Sectional view of the loading pin shown at the various phases of the loading process.

during transfer. The top of the pin is introduced into the probe solutions that are kept in microtiter plates, upon which a small 1-nL droplet is formed on the tip; the pin then touches the gel element surface, and the sample is transferred (**Fig. 4**). A manual version of this procedure is also available, in which the application is carried out with a pipet under a regular microscope *(6)*.

The chemical immobilization of the activated probes to the gel elements is the next step of microchip production. We have been routinely using two methods for immobilization. In the first method, the gel supports contain amino or aldehyde groups allowing coupling with oligonucleotides bearing aldehyde or amino groups, respectively (**Fig. 3**) *(9)*. In the second method, the polyacrylamide gel matrix is activated by introducing hydrazide groups that interact with the 3'-dialdehyde termini of activated oligonucleotides. The disadvantage of this method is that the hydrazide chemistry does not provide sufficient stability of attachment in repeated hybridization experiments.

2.4. Preparation of Target

The target molecules need to be labeled to allow monitoring of the interaction reactions. Although we have used radioactive labeling in the past, our present technology is based on labeling with fluorescent dyes. The advantages of using fluorescence detection are many, including the possibility to monitor processes in real time, high spatial resolution, and lack of radiation hazards. Several criteria need to be met by a labeling procedure:

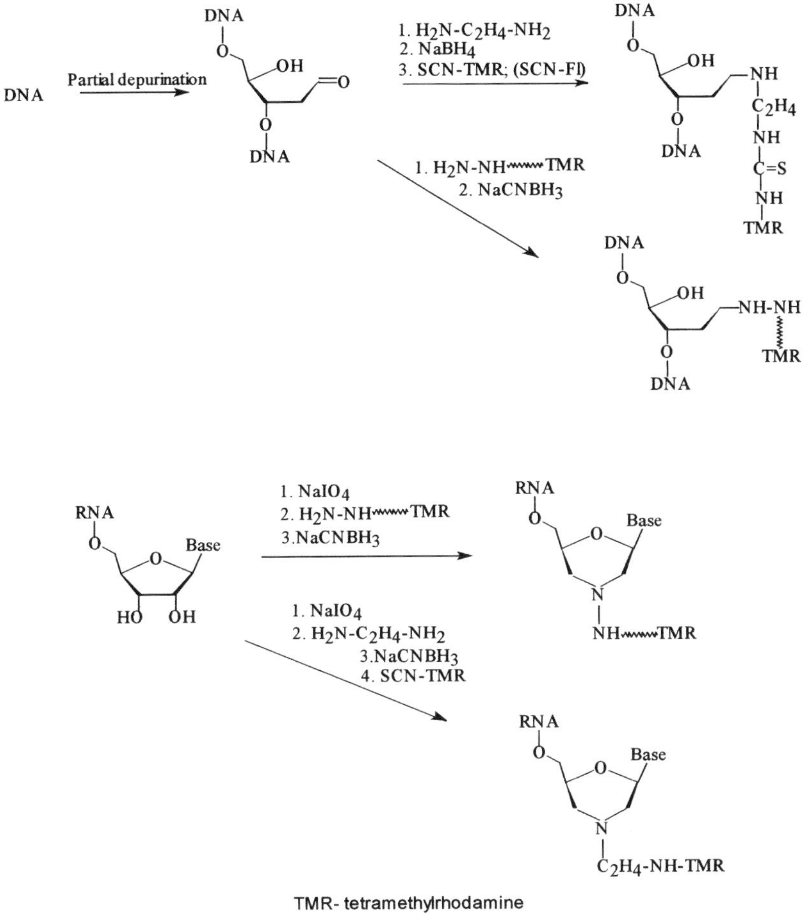

Fig. 5. Chemistry of the procedures used for labeling of DNA or RNA targets. TMR, tetramethylrhodamine.

1. It should be simple, fast, and inexpensive.
2. It should be applicable to both RNA and DNA targets.
3. It should be compatible with the fragmentation often required to decrease secondary structure formation (*see* below).
4. It should allow incorporation of one label into one fragment to ensure proper quantitation of the hybridization intensity.
5. It should allow coupling of multiple dyes.

We have developed a useful procedure (*10*) that is based on the introduction of aldehyde groups by partial depurination of DNA or oxidation of the 3'-terminal ribonucleoside in RNA by sodium periodate (**Fig. 5**). Fluorescent dyes with attached hydrazine group are efficiently coupled with the aldehyde groups, and the bond is stabilized by reduction. An alternative procedure *(10)* uses ethylenediamine splitting of the DNA at the depurinated sites, stabilization of the aldimine bond by reduction, and coupling of the introduced primary amine groups with isothiocyanate or succinimide derivatives of the dyes. New methods for efficient simultaneous radical-based fragmentation and labeling are also being developed. Other published procedures based on reaction of abasic sites in DNA with fluorescent labels containing an oxyamino group *(12)* can also be used in target preparation.

2.5. Performing Hybridization: Some Theoretical Considerations

The basic principle underlying the use of oligonucleotide and DNA biochips is the discrimination between perfect and mismatched duplexes. The efficiency of discrimination depends on a complex set of parameters *(13,14)*, such as the position of the mismatch in the probe, the length of the probe, its AT-content, and the hybridization conditions. Thus, e.g., central mismatches are easier to detect than terminal ones, and shorter probes allow easier match/mismatch discrimination, although the overall duplex stability decreases as the length of the oligomer decreases, which may lead to prohibitory low hybridization signals with shorter probes.

Significant differences may exist in duplex stability depending on the AT content of the analyzed duplexes. This difference stems from the rather large difference in the stability of the AT and CG base pair (two vs three hydrogen bonds). The situation is further complicated because the stability is also sequence dependent: duplexes of the same overall AT content may have different stabilities depending on the mutual disposition of the nucleotides. Several approaches have been used to equalize the thermal stability of duplexes of differing base compositions, including using probes of different lengths, and performing the hybridization in the presence of tetramethylammonium chloride, or betaine *(15)*.

If the technology allows monitoring of melting curves of duplexes formed with individual probe *(16)*, then it is possible to optimize the reaction conditions in order to improve the discrimination of perfect/mismatched duplexes. Note that the melting temperature, T_m, of duplexes formed with matrix-immobilized oligonucleotides is a function of the concentration of the test sample (and is independent of the concentration of the immobilized species). The higher the concentration of the test sample, the more thermodynamically favorable the binding, and, hence, the higher the T_m. When melting is carried out in excess of target molecules, i.e., under conditions of saturation of all binding sites in the gel pad at low temperature, then no match/mismatch discrimination is possible at this temperature. Raising the temperature to the T_m

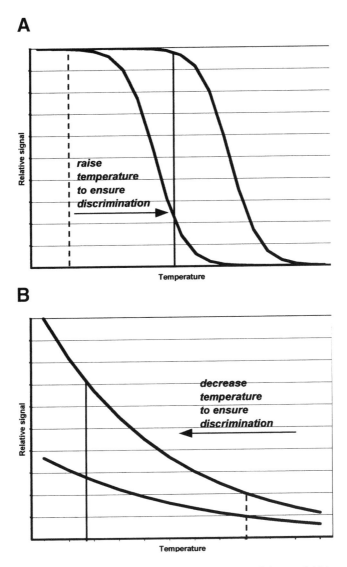

Fig. 6. Melting curves of duplexes formed under conditions of (**A**) target excess and (**B**) low concentration of target. The schemes illustrate the importance of correct temperature selection to ensure discrimination between duplexes of different thermo-dynamic stabilities. The temperatures that will not allow discrimination are represented by dashed lines, whereas the temperatures at which discrimination will be easily achieved are represented by the solid lines.

will ensure the discrimination (**Fig. 6**). On the other hand, if hybridization is carried out at low concentration of the target, so that even the low temperature will not lead to saturation of the probe, it will be necessary to decrease the temperature to enhance discrimination. In such a case, both the perfect and the mismatched signals and the difference between them will be increased.

The previous discussion refers to thermodynamic equilibrium differences in the stability of the perfect and mismatched duplexes, and will be valid only under equilibrium hybridization conditions. Discrimination may, however, be achieved through alternative, kinetic differences. For instance, posthybridization washes can drastically reduce the mismatched signals, almost without affecting the perfect duplexes, in view of the faster dissociation of the mismatched ones.

An interesting twist in approaching the AT content problem came from the unexpected experimental observation that if the oligonucleotide probes are immobilized in three-dimensional gel pads, the apparent dissociation temperature, T_d (defined as the temperature at which the initial hybridization signal decreases 10-fold during step wise heating, posthybridization washing), is actually dependent on the concentration of the immobilized oligonucleotides (5). (For the usual first-order dissociation reaction in solution, the kinetics should be probe concentration-independent.) Our analysis suggests that the diffusion of the dissociated test molecules through the gel pad is retarded by encountering and reversibly binding to other probes immobilized at high density within the gel pad. This retarded diffusion is then probe concentration dependent and creates the apparent probe concentration dependence of the dissociation as a whole. This experimental observation was used to derive an algorithm that allows the design of "normalized" oligonucleotide matrices in which a higher concentration of AT-rich and lower concentration of GC-rich immobilized oligonucleotides can be used to equalize apparent dissociation temperatures of duplexes differing in their AT content, thus facilitating true match/mismatch discrimination.

Finally, we need to note the possibility of using chemically modified nucleotides to improve the discrimination. Examples of such use have been reported (17), and our own unpublished experiments clearly demonstrate the feasibility of such an approach.

Another issue that requires careful consideration is the effect secondary structures in single-stranded nucleic acids may have on the hybridization. The same conditions that favor duplex formation between the immobilized probe and the target will also favor intrastrand duplexing, thus making the target sequence inaccessible for intermolecular complex formation. The use of peptide nucleic acids as probes, rather than standard oligonucleotides, has been described (18) to circumvent this obstacle. We have chosen to prevent the formation of stable secondary structures in the target molecules by performing random fragmentation and fluorescent labeling of the targets under conditions in which the duplexes are melted, e.g., by high temperature. The use of such fragmented targets for hybridization is efficient and produces signals of high intensity.

In summary, even this brief description makes it clear that the design of the biochip and the hybridization conditions should be carefully selected to give unambiguous and reproducible results.

2.6. On-Chip Amplification Reactions

Use of biochip technology will be greatly broadened if on-chip amplification of the hybridization reaction could be performed. This is a highly desirable feature in cases when the nucleic acid of interest presents only a relatively small portion of the molecular population applied on the chip, e.g., when one is dealing with single-copy genes or with mRNAs of low abundance. With this in mind, we are developing methods for on-chip amplification.

In a single-base extension approach *(19)*, a primer is hybridized to DNA and extended with DNA polymerase by a dideoxyribonucleoside triphosphate that matches the nucleotide at a polymorphic site. In our method *(20)*, we perform the single-base extension reaction isothermally, at elevated temperatures, in the presence of each of the four fluorescently labeled ddNTPs (**Fig. 7A**). Performing the extension at a temperature above the melting temperature of the duplex between the DNA and the immobilized primer allows rapid association/dissociation of the target DNA. Thus, the same DNA molecule interacts in succession with many individual primers, leading to amplification of the signal in each individual gel pad. In an alternative procedure, the biochip contains four immobilized primers that differ at the 3' end by carrying one of the four possible nucleotides, matching the polymorphic site (**Fig. 7B**). In this case, extension of the primer will occur only in the gel pads where the primer forms a perfect duplex with the target DNA. Both procedures were applied to the identification of β-globin gene mutations in β-thalassemia patients, and to the detection of anthrax toxin gene *(20)*.

We are also in the process of performing bona fide PCRs directly on the chip, with high expectations of success. In principle, the capability of the chip to perform individual PCRs in individual gel pads depends on the possibility of isolating each pad from its neighbors, which is trivial with our technology but may present insurmountable obstacles in other available chip platforms.

2.7. Readout

For the analysis of the hybridization results obtained with fluorescently labeled target molecules, we use instrumentation constructed in collaboration with the State Optical Institute in S. Petersburg, Russia (**Fig. 8**). The instruments are based on research-quality fluorescence microscopes employing cus-

A   **B**

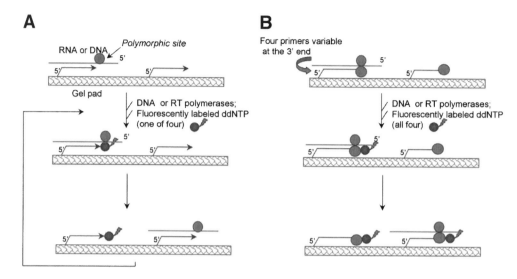

Fig. 7. Schemes illustrating the principle of single-nucleotide extension. (**A**) Multibase extension. (**B**) Multiprimer extension. The extension reactions are performed isothermically, at high temperature, to ensure "jumping" of the target molecule from one immobilized primer to another. For further details, *see* text.

tom-designed, wide-field, high-aperture, large-distance optics, and a high-pressure mercury lamp as a light source for epiillumination. Interchangeable filter sets allow the use of fluorescein, Texas Red, and tetramethylrhodamine derivatives as labeling dyes. The instruments are equipped with a controlled-temperature sample table, which allows changing the temperature in the range from –10 to +60°C in the chip-containing reaction chamber during the course of the experiment. The position of the thermotable can also be changed in a stepwise manner, to allow two-dimensional movement of the sample and analysis of different fields of view. A cooled charge-coupled device (CCD) camera is used to record the light signals from the chip, which are then fed into the analyzing computer program for quantitative evaluation of the hybridization signals over the entire chip.

At present, we are using four different variants of the microscope-based reading instrumentation that differ in their performance level. An important advantage of these devices is that they allow real-time monitoring of the changes in hybridization signal in each individual gel pad under a wide variety of experimental conditions. Most important, they allow monitoring of melting curves, which, in some cases, may be crucial in the proper match/mismatch discrimination. Such instrument capabilities are also important in studies of

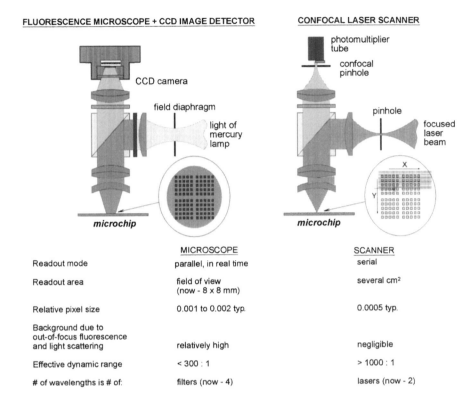

Fig. 8. Types of reading instrumentation, fluorescence microscope, and laser scanner and their basic characteristics.

the specificity of binding of sequence-specific ligands to single- and double-stranded DNA, because such specific binding raises the melting temperature of the ligand-bound duplexes to a measurable and interpretable degree. The feasibility of such an approach has recently been demonstrated in studies of Hoechst binding to DNA *(21)*.

Although the most widely used in our current experimental practice, the conventional imaging fluorescence microscopy is not the only approach to microchip readout that is under development in our group. In many cases, when parallel measurements of gel pad signals are essential because of possible data loss, a more cost-effective solution of the readout problem can be offered using laser-scanning platforms. Because of inherently low background and excellent uniformity of the fluorescence excitation and detection, microchip scanners are especially well suited for precise quantitative measurements of signals varying over the range of three or even more orders of magnitude. However, all commercially available scanners are closed-architecture instruments optimized with the surface-immobilization microchips in mind. Typically, they lack such a useful feature as temperature control of the sample table and employ an objective lens with a working distance too small to accommodate microchips packaged in a hybridization cell.

To meet the specific requirements of gene expression studies and cost-sensitive diagnostic applications, we have recently developed a laser scanner of unique, nonimaging design that makes use of the well-defined geometry of the gel-based microchips. The scanner employs a 2-mW HeNe laser as an excitation source and a low-noise PIN photodiode as a detector. The laser wavelength (594 nm) almost perfectly matches the absorption band of Texas Red. The numerical aperture of the miniature objective lens is 0.62. Yet, its working distance (approx 3 mm) is long enough for scanning packaged microchips. A microchip is mounted on a stationary controlled-temperature sample table of a design similar to that used in our fluorescence microscopes. All parameters of the scanning, data visualization, and processing are set up via the host computer. The hybridization pattern can be stored in a file either in the raw-data format or as an array of integral fluorescence intensities calculated on-line per each gel pad. Using a Texas Red dilution series microchip, we determined the detection threshold (3 σ) of the scanner to be approx 2 amol of Texas Red per gel pad, with a linear dynamic range being up to three orders of magnitude in terms of integral signal intensities. These characteristics are close to those of a commercial ScanArray 300 scanner (General Scanning) that we use for routine microchip inspection.

2.8. Informatics

The digitized images of hybridization patterns obtained with the help of the CCD camera are further treated with the help of specialized software. This treatment includes automatic image analysis that determines the localization of the rows and columns of the matrix gel pads and their centers. For each element that contains a large number of pixels, the program calculates the total intensity of the hybridization signal. The program allows, if the need arises, filtering of the image in order to remove any noise coming from fluorescent impurities (e.g., dust particles) in the gel. The computer then performs, based on the calculated intensities of all gel pads on the chip image and stored information on standard image patterns, recognition operations. Such operations help the investigator obtain the final results in a user-friendly format.

3. Types of MAGIChips and Examples of Application

3.1. Oligonucleotide Chip

3.1.1. Customized Oligonucleotide Arrays

Customized oligonucleotide biochips are designed to interrogate test samples of known nucleotide sequences. Such sequences may be those of known genes in cases when one is interested in their expression levels under specified conditions, or sequences of genes that are known or expected to contain single-nucleotide mutations, or to be polymorphic in a given population of individuals. In all these cases, the proper choice of oligonucleotide probes to be immobilized is of crucial importance to the success of the assay.

The choice of the oligonucleotide probes depends on the particular application, but there are certain basic considerations common to all cases. A basic requirement is to minimize the number of probes to be immobilized because such minimization may lead to a simplified design of the chip, which, in turn, makes its production cheaper and its use more convenient. A smaller chip will be easier to handle and read and will require a smaller amount of test sample. On the other hand, the set of selected probes must be big enough to allow unambiguous identification of the test samples; ambiguity might arise in view of the inherent to the procedure variations in the intensity of the hybridization signal in individual gel pads.

With these general requirements, a set of potential probes is created for each interrogated sequence that form perfect duplexes with that sequence. Then, some of the potential probes that may create ambiguities in the interpretation of the hybridization pattern are excluded from this list based on the AT vs GC content, and the propensity to form hairpins and other types of stable secondary structures that may drastically affect the intensity of hybridization. At a final stage of selection, all members of the shortened list are checked for their uniqueness: the probes should not form duplexes similar in their stability to the perfect duplexes with any material present in the mixture applied to the chip. This round of selection thus compares the sequence of each selected probe to the sequences in the set of test samples that are supposed to be distinguished from each other on the chip. The comparison of duplex stabilities takes into consideration not only the number of existing mismatches but also their location with respect to the probe because the stabilities of the perfect and mismatched duplexes may differ insignificantly if the mismatched nucleotide is close to the end of the probe, especially in cases of longer probes.

Examples of successful applications of customized oligonucleotide chips include detection of β-thalassemia mutation in patients *(11,20,22)*, gene identification *(23)*, allele-type identification in the human HLA DQA1 locus *(24)*, identification of polymorphic base substitutions in patients with neurological disorders *(25)*, and identification of and discrimination among closely related bacterial species *(26)*. For the diagnostics of β-thalassemia mutations, a simple chip was designed that contained six probes corresponding to different β-thalassemia genotypes and hybridized with PCR-amplified DNA from healthy humans and patients *(11)*. The hybridization results showed the expected significant differences in signal intensity between matched and mismatched duplexes, thus allowing reliable identification of both homozygous and heterozygous mutations.

For the bacterial biochip, oligonucleotides complementary to the small-sub-unit rRNA sequences were immobilized on the chip and hybridized with either DNA or RNA forms of the target sequences from nitrifying bacteria *(26)*. This biochip successfully identified the tested microorganisms. In addition, the system was used to verify the utility of varying the concentrations of the immobilized oligonucleotides to normalize hybridization signals, and to demonstrate the use of multicolor detection for simultaneous hybridization with multiple targets labeled with different fluorescent dyes. In another application, chips based on sequences from 16S rRNA were successfully used to discriminate among closely related pathogenic and nonpathogenic bacterial species.

3.1.2. Generic Oligonucleotide Arrays

A more universal array that can be used to interrogate any target sequence is the so-called generic array that uses complete sets of small oligonucleotides of a given length. Such arrays were originally proposed for *de novo* DNA sequencing *(3,27,28)*, but some intrinsic problems have, for the time being, hampered their implementation to such sequencing. These are mainly connected to sequence reconstruction ambiguities stemming from the presence of repeats along the DNA, and because relatively short stretches of nucleotides (of the length of the immobilized probes) can be found in many positions of a complex DNA sample. These restrictions severely limit the length of the DNA fragment that can be successfully decoded on short oligonucleotide arrays. Although the length and complexity of the readable DNA can be increased by using arrays of longer oligonucleotides, the gain in length is expected to grow linearly with the length of the probes *(29)*, whereas the number of probes in a complete set of *n*-mers increases exponentially with probe length (4^n). Thus, a complete array of 6-mers contains 4096 members, of 7-mers, 16,384 members; of 8-mers, 65,536; of 9-mers, 262,144; and so on. The production of such arrays still poses a practical challenge.

We have suggested that the use of contiguous stacking hybridization *(4)* can largely overcome the need of such impractically large arrays. In this approach, the initial hybridization of the target DNA with the array containing the full set of oligonucleotides of fixed length *L* is followed by additional multiple rounds of hybridizations with fluorescently labeled oligonucleotides of length *l* (**Fig. 9**). These labeled oligonucleotides will form extended duplexes with the target DNA strand owing to the strong stacking interactions between the terminal bases of the existing DNA duplex with the immobilized probe and of the otherwise unstable duplex with the short labeled probe (**Fig. 9**). The stacking interactions stabilize the DNA duplex even in the absence of a phosphodiester bond or a phosphate group. The stacked *l*-mers can also be ligated to the probe (*[30]*; for on-chip ligation, *see* **ref. 23**). Theoretical calculations have demonstrated

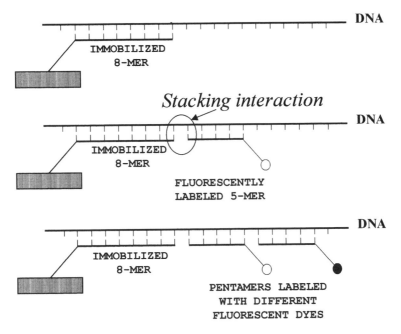

Fig. 9. Principle of contiguous stacking hybridization.

that contiguous stacking hybridization considerably increases the resolution power of the matrix, which approaches the power of $(L + l)$-matrix *(29)*. An algorithm has been developed that allows the minimization of the number of additional hybridization steps, by assembling sets of l-probes to be added together at each hybridization round *(29)*.

Recently, an ingenious approach that uses the ligation of DNA targets to arrays containing duplex probes with 5'-mer overhangs has been reported by Affymetrix *(31)*. Even this approach allowed only for unambiguous sequence verification, not *de novo* sequencing, of relatively long targets (1200 bp). Note that although ligation certainly expands the resolution power of the chip, such chips cannot be used more than once. This limitation will be especially undesirable with more complex, expensive arrays.

In principle, contiguous stacking hybridization may be used with customized oligonucleotide chips too *(32)*. In this work, an alternative hybridization strategy uses sets of labeled "stacking" oligonucleotides, each containing a "discriminating" base at one position and a universal base or a mixture of all four bases at all other positions. Such an approach allows 1024 rounds of hybridization with all possible 5-mers to be replaced by only five rounds of hybridization. This study has demonstrated that the 5-mers are stabilized in duplexes even with weak stacking bases and that mismatches in any of the five positions of the 5-mers drastically destabilize the stacked duplexes. The desta-bilizing effect of mismatches in the 5-mers was also convincingly shown in a study of point mutations of the β-globin gene associated with β-thalassemia *(11)*. Because the destabilizing effect of mismatches increases dramatically upon decreasing the length of the duplex (i.e., it is much stronger in 5-mers compared with, e.g., 8-mers) the use of contiguous stacking hybridization with short additional probes will be much more advantageous in mismatch discrimi-nation than the conventional one-probe procedure.

3.2. cDNA Microarrays

In cDNA microarrays, the immobilized probes are individual cDNAs obtained by reverse transcription on mRNA populations extracted from cells in different physiological and developmental states. cDNA arrays have been widely used to study gene expression *(33–35)*. A potential problem with cDNA arrays on the MAGIChip could be connected to the difficulty of introducing and evenly distributing long molecules into the gel pads. To facilitate diffusion of longer cDNAs, we are successfully developing polyacrylamide gels of vari-ous compositions that contain larger average pore sizes. Such gels will also be used in protein chips, where it may be necessary to immobilize rather bulky protein molecules within the gel pads (*see* **Subheading 3.3.**). An even more practical solution to the diffusion problem lies in randomly fragmenting the cDNA into relatively small pieces before immobilization.

3.3. Protein MAGIChips

It is our goal to expand the capabilities of our technology by producing pro-tein chips. Such chips should contain different proteins immobilized as probes in a way that preserves their biological activity. The feasibility of producing such gels has been demonstrated *(6,36)*.

One potential limitation with protein chips may stem from the difficulty of diffusion within the gel pads of molecules of high molecular masses. A way to circumvent this limitation is being sought in the production of polyacrylamide gels of larger pore sizes. In principle, larger-pore gels can be prepared by increasing the ratio of the crosslinker N,N'-methylenebisacrylamide (Bis) to acrylamide *(37)*, or by the use of alternative crosslinkers such as N,N'-diallyltartardiamide *(6)*, or a mixture of Bis and N,N'-(1,2-dihydroxyethylene)-bisacrylamide *(36–38)*.

We have tested two protein-immobilization procedures *(36)*. The first is based on activation of the polyacrylamide gel with glutaraldehyde *(39)*. In the second procedure, which is applicable to glycoproteins such as antibodies, the gel is activated by partial substitution of the amide groups with hydrazide groups, and the polysaccharide component of the protein is activated by $NaIO_4$

A **B**

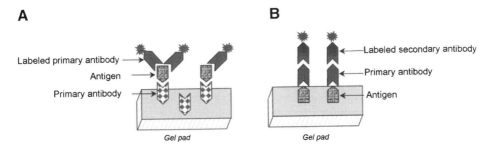

Fig. 10. Indirect approaches for detection of antigen-antibody reactions performed on the chip: the target is unlabeled, and a positive reaction is registered by using labeled molecules that specifically recognize the target. (**A**) The chip contains immobilized primary antibodies to interrogate the sample for the presence of a specific antigen. Detection is through binding of fluorescently labeled primary antibodies that recognize antigenic determinants distinct from the ones recognized in the primary reaction. (**B**) The chip contains immobilized antigen that is recognized by primary antibodies present in the solution. The reaction is detected via binding of secondary antibodies (antibodies to the first antibody), as in classical immunochemical techniques.

oxidation. The reaction between the hydrazide and aldehyde groups efficiently crosslinks the protein to the gel. This procedure is similar to one of the methods used for oligonucleotide immobilization.

Protein microchips preserve the high specificity of molecular recognition reactions observed in solution. For instance, the interaction between antigens and their specific antibody partners may occur on-chip in a variety of experimental setups. Either the antigen or the antibody can be immobilized, and both direct and indirect monitoring reactions can be performed. In the direct methods, one uses target molecules labeled with fluorescent dyes or coupled to enzymes catalyzing color precipitate-forming reactions. In the indirect methods, the target is unlabeled, and the reaction is detected by using a labeled molecule that specifically recognizes the target. Examples include the use of secondary antibodies to detect primary antibodies bound to the immobilized antigen, or the use of labeled primary antibodies to detect antigens bound to immobilized specific antibodies (**Fig. 10**).

Finally, the protein biochip allows the study of enzymatic activity of immobilized enzymes, by overlaying the chip with solutions containing the respective substrates. The reaction is monitored by following the formation of color or fluorescent precipitates; more important, this can be done in real-time experiments. We have been successful in demonstrating the feasibility of this approach with enzymes of different molecular masses: horseradish peroxidase (44 kDa), alkaline phosphatase (140 kDa), and β-D-glucoronidase (290 kDa).

The enzyme biochips present an important future application for combinatorial drug discovery, in view of the possibility of detecting the effect of inhibitors on enzymatic activity *(36)*.

Acknowledgments

We acknowledge the members of our team for help in preparation of the manuscript, in particular Drs. V. Barsky, S. Bavykin , M. Livshitz, Y. Lysov, A. Perov, D. Proudnikov, V. Shick, E. Timofeev, and G. Yershov. We also acknowlege the skillful secretarial help of Felicia King. This work was supported by the U.S. Department of Energy, Office of Health and Environmental Research, under contract no. W-31-109-Eng-38; the Defense Advanced Research Project Agency, under interagency agreement no. AO-E428; CRADA grant no. 9701902 with Motorola and Packard Instrument Company; and by grants no. 558 and 562 of the Russian Human Genome Program.

References

1. (1998) The runners-up. *Science* **282,** 2156–2157.
2. (1999) *Nat. Genet.* **21(Suppl.).**
3. Lysov, Y., Chernyi, A., Balaeff, A., Beattie, K., and Mirzabekov, A. (1994) DNA sequencing by hybridization to oligonucleotide matrix. Calculation of continuous stacking hybridization efficiency. *J. Biomol. Struct. Dyn.* **11,** 797–812.
4. Khrapko, K., Lysov, Y., Khorlin, A., Ivanov, I., Yershov, G., Vasilenko, S., Florentiev, V., and Mirzabekov, A. (1991) A method for DNA sequencing by hybridization with oligonucleotide matrix. *DNA Sequence* **1,** 375–388.
5. Khrapko, K., Lysov, Y., Khorlin, A., Shick, V., Florentiev, V., and Mirzabekov, A. (1989) An oligonucleotide hybridization approach to DNA sequencing. *FEBS Lett.* **256,** 118–122.
6. Guschin, D., Yershov, G., Zaslavsky, A., Gemmell, A., Shick, V., Proudnikov, D., Arenkov, P., and Mirzabekov, A. (1997) Manual manufacturing of oligonucleotide, DNA and protein microchips. *Anal. Biochem.* **250,** 203–211.
7. Vasiliskov, A. V., Timofeev, E. N., Surzhikov, S. A., Drobyshev, A. L., Shick, V. V., and Mirzabekov, A. D. (1999) Fabrication of microarray of gel-immobilized compounds on a chip by copolymerization. *Biotechniques* **27,** 592–606.
8. Proudnikov, D., Timofeev, E., and Mirzabekov, A. (1998) Immobilization of DNA in polyacrylamide gel for manufacturing of DNA and DNA-oligonucleotide microchips. *Anal. Biochem.* **259,** 34–41.
9. Timofeev, E., Kochetkova, S., Mirzabekov, A., and Florentiev, V. (1996) Regioselective immobilization of short oligonucleotides to acrylic copolymer gels. *Nucleic Acids Res.* **24,** 3142–3148.
10. Proudnikov, D. and Mirzabekov, A. (1996) Chemical methods of DNA and RNA fluorescent labeling. *Nucleic Acids Res.* **24,** 4535–4542.
11. Yershov, G., Barsky, V., Belgovskiy, A., Kirillov, E., Kreindlin, E., Ivanov, I., Parinov, S., Guschin, D., Drobyshev, A., Dubiley, S., and Mirzabekov, A. (1996)

DNA analysis and diagnostics on oligonucleotide microchips. *Proc. Natl. Acad. Sci. USA* **93**, 4913–4918.

12. Boturyn, D., Defancq, E., Ducros, V., Fontaine, C., and Lhomme, J. (1997) Quantitative one step derivatization of oligonucleotides by a fluorescent label through abasic site formation. *Nucleosides Nucleotides* **16**, 2069–2077.

13. Livshits, M., Florentiev, V., and Mirzabekov, A. (1994) Dissociation of duplexes formed by hybridization of DNA with gel-immobilized oligonucleotides. *J. Biomol. Struct. Dyn.* **11**, 783–795.

14. Livshits, M. and Mirzabekov, A. (1996) Theoretecal analysis of the kinetics of DNA hybridization with gel-immobilized oligonucleotides. *Biophys. J.* **71**, 2795–2801.

15. Mirzabekov, A. (1994) DNA sequencing by hybridization—a megasequencing method and a diagnostic tool? *Trends Biotech.* **12**, 27–32.

16. Fotin, A. V., Drobyshev, A. L., Proudnikov, D. Y., Perov, A. N., and Mirzabekov, A. D. (1998) Parallel thermodynamic analysis of duplexes on oligodeoxyribonucleotide microchips. *Nucleic Acids Res.* **26**, 1515–1521.

17. Mirzabekov, A. D. (1999) *Biological Agent Detection and Identification DARPA Conference Proceedings,* April 27–30, Santa Fe, NM, pp. 57–67.

18. Marshall, A. and Hodgson, J. (1998) DNA chips: An array of possibilities. *Nature Biotech.* **16**, 27–31.

19. Syvanen, A. C., Aalto-Setala, K., Harju, L., Kontula, K., and Soderlund, H. (1990) A primer-guided nucleotide incorporation assay in the genotyping of apolipoprotein E. *Genomics* **8**, 684–692.

20. Dubiley, S., Kirillov, E., and Mirzabekov, A. (1999) Polymorphism analysis and gene detection by minisequencing on an array of gel-immobilized primers. *Nucleic Acids Res.* **27**, e19.

21. Drobyshev, A. L., Zasedatelev, A. S., Yershov, G. M., and Mirzabekov, A. D. (1999) Massive parallel analysis of DNA Hoechst 33258 binding specificity with a generic oligodeoxyribonucleotide microchip. *Nucleic Acids Res.* **27**, 4100–4105.

22. Drobyshev, A., Mologina, N., Shick, V., Pobedimskaya, D., Yershov, G., and Mirzabekov, A. (1997) Sequence analysis by hybridization with oligonucleotide microchip: identification of beta-thalassemia mutations. *Gene* **188**, 45–52.

23. Dubiley, S., Kirillov, E., Lysov, Y., and Mirzabekov, A. (1997) Fractionation, phosphorylation, and ligation on oligonucleotide microchips to enhance sequencing by hybridization. *Nucleic Acids Res.* **25**, 2259–2265.

24. Shik, V. V., Lebed, Y. B., and Kryukov, G. V. (1998) Identification of HLA D*Q*A1 alleles by the oligonucleotide microchip method. *Mol. Biol.* **32**, 679–688.

25. LaForge, K. S., Shick, V., Sprangler, R., Proudnikov, D., Yuferov, V., Lysov, Y., Mirzabekov, A., and Kreek, M. J. (2000) Detection of single nucleotide polymorphisms of the human mu opioid receptor gene by hybridization or single nucleotide extension on custom oligonucleotide gelpad microchips: Potential in studies of addiction. *Am. J. Med. Genet.*, in press.

26. Guschin, D., Mobarry, B., Proudnikov, D., Stahl, D., Rittman, B., and Mirzabekov, A. (1997) Oligonucleotide microchips as genosensors for determi-

native and environmental studies in microbiology. *Appl. Environ. Microbiol.* **63,** 2397–2402.

27. Bains, W. and Smith, G.C. (1988) A novel method for nucleic acid sequence determination. *J. Theor. Biol.* **135,** 303–307.

28. Drmanac, R., Labat, I., Brukner, I., and Crkvenjakov, R. (1989) Sequencing of megabase plus DNA by hybridization: Theory of the method. *Genomics* **4,** 114–128.

29. Lysov, Y., Florentiev, V., Khorlin, A., Khrapko, K., Shick, V., and Mirzabekov, A. (1988) A new method to determine the nucleotide sequence by hybridizing DNA with oligonucleotides. *Proc. Natl. Acad. Sci. USSR* **303,** 1508–1511.

30. Broude, N. E., Sano, T., Smith, C. L., and Cantor, C. R. (1994) Enhanced DNA sequencing by hybridization. *Proc. Natl. Acad. Sci. USA* **91,** 3072–3076.

31. Gunderson, K. L., Huang, X. C., Morris, M. S., Lipshutz, R. J., Lockhart, D. J., and Chee, M. S. (1998) Mutation detection by ligation to complete *n*-mer DNA arrays. *Genome Res.* **8,** 1142–1153.

32. Parinov, S., Barsky, V., Yershov, G., Kirillov, E., Timofeev, E., Belgovskiy, A., and Mirzabekov, A. (1996) DNA sequencing by hybridization to microchip octa- and decanucleotides extended by stacked pentanucleotides. *Nucleic Acids Res.* **24,** 2998–3004.

33. Schena, M., Shalon, D., Davis, R. W., and Brown, P. O. (1995) Quantitative monitoring of gene expression patterns with a complementary DNA microarray. *Science* **270,** 467–470.

34. DeRisi, J., Penland, L., Brown, P. O., Bittner, M. L., Meltzer, P. S., Ray, M., Chen, Y., Su, Y. A., and Trent, J. M. (1996) Use of a cDNA microarray to analyse gene expression patterns in human cancer. *Nat. Genet.* **14,** 457–460.

35. Duggan, D. J., Bittner, M., Chen, Y., Meltzer, P., and Trent, J. M. (1999) Expression profiling using cDNA microarrays. *Nat. Genet.* **21,** 10–14.

36. Arenkov, P., Kukhtin, A., Gemmell, A., Voloschuk, S., Chupeeva, V., and Mirzabekov, A. (2000) Protein microchips: use for immunoassay and enzymatic reactions. *Anal. Biochem.* **278,** 123–131.

37. Righetti, P. G., Brost, B. C. W., and Snyder, R. S. (1981) On the limiting pore size of hydrophilic hydrogels for electrophoresis and isoelectric focusing. *J. Biochem. Biophys. Methods* **4,** 347–363.

38. O'Connell, P. B. H. and Brady, C. J. (1976) Polyacrylamide gels with modified cross-linkages. *Anal. Biochem.* **76,** 63–73.

39. Hermanson, G. T. (1996) *Bioconjugate Techniques,* Academic, San Diego, CA.

3

Sequencing by Hybridization Arrays

Radoje Drmanac and Snezana Drmanac

1. Introduction

By determining an organism's DNA sequence, researchers can obtain critical information about its development and physiology, its taxonomic relations, and its susceptibility to disease. There are three distinct methods of acquiring DNA sequence information: sequence-specific DNA degradation, synthesis, and/or separation; sequence-specific DNA hybridization with oligonucleotide probes; and nucleotide chain visualization. This chapter focuses on the second of these processes: the use of sequence-specific hybridization of oligonucleotide probes of known sequence to determine primary DNA structure. Refined over the past decade, such sequencing-by-hybridization (SBH) methods have become important tools in the field of genomics research.

SBH methods take advantage of one of the fundamental chemical processes of life—the molecular specificity of pairing that occurs between complementary DNA strands. Oligonucleotide probes of known sequence are tested for complementary pairing (e.g., hybridization) against a DNA target. The process, which in some respects is similar to a keyword search of text by an Internet browser, is designed to identify matching sequence strings within the target DNA. In one scenario, software programs are then used to assemble the target sequence by ordering the set of overlapping, high-scoring probes.

SBH techniques may involve pairwise hybridizations of thousands of oligonucleotide probes with each DNA sample. As a result, such procedures are ideally suited to microarray chemistry, with its advantages of miniaturization and parallel processing of hybridization reactions. Either DNA samples (Format 1 SBH) or oligonucleotide probes (Format 2 SBH) may be attached to a solid support to form an array, with the other member of the hybridization pair added in solution on each array spot. In Format 3 SBH procedures, combinato-

From: *Methods in Molecular Biology, vol. 170: DNA Arrays: Methods and Protocols*
Edited by: J. B. Rampal © Humana Press Inc., Totowa, NJ

rial ligation of an attached and a labeled short probe that bind to precisely adjacent positions in a target DNA allows the scoring of potentially millions of longer sequence strings within the target DNA. We describe here the use of complete oligonucleotide probe sets to obtain the full sequence of any DNA sample of appropriate length using SBH Formats 1 and 3.

1.1. Short History of Hybridization

In its natural state, DNA consists of a helical duplex, in which two complementary single-stranded DNA chains specifically intertwine with one another. Each of the purine DNA bases in one strand binds a specific pyrimidine partner in the complementary strand, adenine to thymine, and guanine to cytosine. The sequence of each DNA strand thus specifically defines the sequence of its complementary partner.

The tendency of complementary DNA strands to pair or renature after melting was first observed in 1960 by Doty et al. *(1)*. Their work initiated the development of many DNA hybridization techniques in molecular biology, such as Southern blots *(2)*. Oligonucleotide probes shorter than about 25 bases have proved to be more accurate sequence-specific detectors than longer, natural probes. In 1979, Wallace et al. *(3)* demonstrated that oligonucleotide probes as short as 11 bases in length can be used to discriminate between perfect probe-target duplexes and those containing a single internal base mismatch. In 1989, Saiki et al. *(4)* described a "reverse dot blot" method of testing patient DNA samples for known mutations based on specificity of oligonucleotide hybridization. In this method, patient samples prepared by polymerase chain reaction (PCR) are tested for hybridization against an array of oligonucleotides designed to complement normal sequences and known mutations. In 1986, Breslauer et al. *(5)* demonstrated methods of predicting sequence-dependent oligonucleotide hybrid stability.

1.2. Sequencing by Hybridization

The method of SBH was first disclosed in a patent application filed by Drmanac and Crkvenjakov *(6)* in 1987. They described a method for sequencing a DNA molecule by obtaining hybridization data from a probe set containing overlapping probes under match-specific conditions. The researchers showed that complex DNA sequences could be reconstructed based on the relationship between the length and number of probes to the length and number of clone fragments from the DNA being sequenced *(6,7)*.

Other research groups subsequently presented a variety of formats and analyses using the SBH process. Bains and Smith *(8)* examined the process of sequence reconstruction using tetramers and gapped hexamer probes. In a 1988 patent application, Southern et al. *(9)* proposed a method for combinatorial

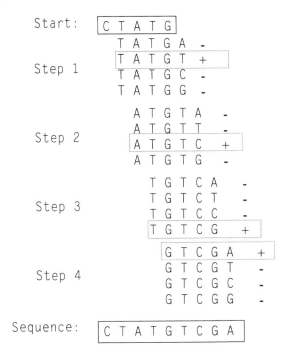

Fig. 1. Sequence assembly using overlapping oligonucleotide probes. Base-by-base sequence extension is determined by which of four possible probes (which differ only at the last base) hybridizes the target DNA. Mismatched probes hybridize inefficiently with the target and are removed by selective washing or other means of chemical discrimination. The positive probe from step 1 is used to define the four possible probes for assembly step 2 and so on. Each base is repeatedly "read" by multiple overlapping probes, minimizing the impact of experimental error.

in situ synthesis of large oligonucleotide arrays for mutation detection or complete sequencing. Lysov et al. *(10)* proposed using an array of physically attached oligonucleotides for SBH analysis. The high accuracy of the SBH method was demonstrated in a blind experiment in which 330 bases of DNA in three samples were sequenced without a single error *(11)*.

Figure 1 shows schematically how sets of overlapping probes are used to assemble an unknown DNA using a simple algorithm. Arbitrarily starting with a first positive probe, a set of four probes (one for each possible base) is used to read the next base position. Assuming no repeat sequences within the target, only one of these four probes will match the appropriate base, whereas the other three will form hybrids with a mismatch at the end base. Mismatched probes either bind poorly (form less stable duplexes with the target DNA) or their hybrids are denatured during selective washing *(3,12)*; full-match hybrids are retained and detected. The correct base is thus determined by which of the four oligonucleotide probes binds the target DNA. This positive probe in turn determines the next set of four probes used to decipher the next base. Each base is "read" by multiple probes, since there is a high degree of overlap among the positive probes. Using experimentally determined hybridization signals for all probes corresponding to a target DNA, sequence assembly may start from any probe and proceed in both directions.

The use of overlapping probes allows researchers to determine sequences within a target DNA that are longer than any of the individual overlapping probes. This assembly principle provides a unique advantage of the SBH method: in contrast to other DNA sequencing methods, SBH allows an indirect assignment of which of four bases is present at a given target position without experimentally interrogating each physical position base by base. As a result, SBH permits unprecedented miniaturization and parallelism of the DNA sequencing process, and has the potential to analyze complex DNA targets, including whole genomes, in a single reaction. In addition, the use of redundant, overlapping probes provides accurate base assignments even when false-positive or false-negative probe scoring occurs due to experimental errors.

2. SBH on Arrays of DNA Samples (Format 1 SBH)
2.1. Principles and Advantages of Format 1 SBH

To conveniently handle large numbers of hybridization reactions, SBH experiments are organized in arrays that may contain tens of thousands of individual probes or DNA samples. Such arrays confer significant advantages by allowing rapid parallel processing of tens of thousands of probe-DNA pairs. Arrays also allow significant process miniaturization, reducing space requirements and the amounts of sample, probe, and reagents needed for each experiment.

In Format 1 SBH, large numbers of DNA samples can be processed simultaneously using arrays containing more than 50,000 spots *(13)*. The DNA templates, typically in the range of 200–2000 bases, are usually prepared by PCR and arrayed without any further purification or treatment. Nylon membranes are frequently used as a solid support for such DNA arrays. These membranes consist of a mesh of fibers, allowing three-dimensional binding via pseudo-covalent bonds. Subsequent ultraviolet crosslinking of thymine bases produces covalent interstrand bonding, leading to retention of a higher proportion of the sample DNA strands. Approximately 60–90% of the initially bound DNA is retained, even after repeated cycles of hybridization and washing. Frequent bonding (every 30–50 bases, on average) minimizes target-to-target chain pairing that might interfere with short probe hybridization. The Format 1 SBH process has the simplest sample preparation procedure and represents the most cost-effective way to analyze large numbers of DNA samples.

For complete Format 1 sequencing, a full set of labeled probes (e.g., 16,384 7-mers) may be scored. **Figure 2** shows a sophisticated sample array designed to handle large numbers of probes in parallel, consisting of replica unit arrays containing 64 test and control DNA spots. These small arrays are arranged in larger arrays consisting of 384 unit arrays on a single nylon membrane. Each unit array is exposed to an individual labeled 7-mer probe and then scored for positive hybridization at each DNA spot. By producing thousands of such unit arrays, researchers can quickly obtain hybridization information for large sets of oligomer probes. Using this system and a complete set of 7-mer probes in a blind experiment, a number of homozygote and heterozygote mutations in the *p53* gene were determined with 100% accuracy *(14)*. A suitable DNA read length for SBH experiments using 7-mer probes is up to 2000 bases.

2.2. Discovery of Mutation and Polymorphism Using 7-mer Probes

Format 1 SBH may be used to discover mutations and polymorphisms in a gene of interest. (The materials and experimental protocols involved in a typical Format 1 SBH are described in Chapter 11). First, sample amplicons are prepared by PCR using genomic DNA of test individuals, and a control DNA with a known (reference) sequence for the amplicon of interest is prepared by PCR or chemical synthesis. A subset of probes (1000–2000 per kb of DNA) that fully complements the reference sequence is automatically retrieved from a stock of all possible probes of a given length, and then hybridized against control and patient samples to detect any previously known or novel mutation that may be present. When a mutation is present, oligomer probes corresponding to that region do not bind the test sample owing to the sequence mismatch between probe and DNA. In fact, many overlapping probes are affected at each mutation site, and the low percentage of positive probes at such sites reflects the high probability that the sequence of the test sample differs from that of the control DNA.

The selected probe subset approach is capable of detecting the existence of any sequence change, but it does not determine the nature of the change. This is not a particular limitation in large-scale searches for new polymorphisms. In a 1000-bp amplicon, it is expected that roughly 10–20 polymorphic sites will show >1% frequency of a secondary allele. This means that of 500–1000 individuals analyzed, as few as 10 have to be tested with a larger set of probes to make base assignments at the discovered polymorphic sites.

Researchers at Hyseq, in collaboration with the University of California at San Francisco, are using Format 1 SBH to detect single nucleotide polymorphisms or other sequence variants associated with cardiovascular disease. Selected sets of 7-mer probes corresponding to sequences of candidate genes are used to screen hundreds of patient samples for polymorphisms. **Figure 3** shows

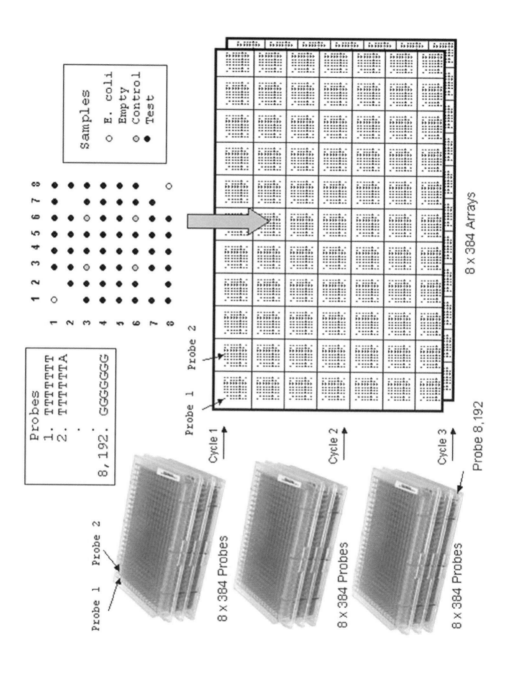

44

polymorphism detection curves for a segment of the *apoB* gene, which is associated with bloodstream lipid metabolism. Such polymorphism detection curves, generated from probe binding data for each PCR amplicon, show the frequency or strength of positive probes for each base position for each patient DNA sample. Probes are deemed "positive" if their hybridization signals are above background and control sample signal thresholds.

A homozygotic polymorphism in a target gene affects a total of 14 overlapping 7-mer probes (seven from each strand). This situation results in a very low positive probe frequency at the polymorphic site, as seen in **Fig. 3A** as the large dip at base position 505 in the otherwise almost flat probe frequency line. A deflection such as this one, with a positive probe frequency of <0.4, indicates a homozygotic mutation in the sample. Heterozygotic mutations or polymorphisms affect only one chromosome (e.g., half of the available template DNA), and hence all probes will score positive, but at one half of the intensity of the control sample. As seen in **Fig. 3B**, this results in a deflection in the heterozygote detection curve, which drops to between 0.5 and 0.8. Slight dips that do not drop below a threshold of 0.8 are generally regarded as experimental noise. Redundant reads (with 14 overlapped probes per base) provide an estimated hundredfold higher accuracy than usually obtained in single-pass gel reactions *(14)*.

3. SBH by Combining Arrayed and Labeled Probes: Principles and Advantages of Format 3 SBH

Format 3 SBH involves the use of DNA ligase or other agents to link together short oligonucleotides under specific hybridization conditions to create longer, support-bound probes *(15)*. (The materials and experimental protocols involved in a typical Format 3 SBH experiment are described in Chapter 11.) This process is illustrated in **Fig. 4**. A short support-bound probe (in **Fig. 4**, an 8-mer) hybridizes to the target DNA in the presence of DNA ligase and another labeled 8-mer probe. If the two probes hybridize to the target DNA at precisely adjacent positions, they are covalently linked by a DNA coupling agent (e.g., ligase) to create a labeled, support-bound 16-mer probe. If either probe fails to hybridize to the target DNA, or the two probes bind at nonadjacent sites, the labeled probe and target DNA are washed off. The high accuracy of this pro-

---

Fig. 2. (*opposite page*) Scoring thousands of labeled probes on arrays of unit sample arrays. The schema shows 8 arrays of 384 replica unit arrays. Each replica unit array, which contains 64 DNA spots and controls, is exposed to a single labeled 7-mer probe and then scored for hybridization at each position within the array. Probes are stored in 384-well plates to match the number of unit arrays. All 16,384 7-mer probes may be scored in three hybridization cycles.

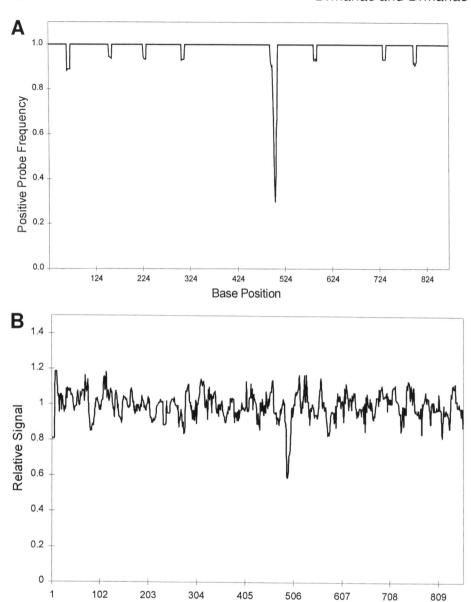

Fig. 3. Polymorphism detection plots. (**A**) Positive probe frequency of a sample with a homozygous A–G mutation at position 505 in an 899-bp amplicon from the *ApoB* gene. This base change results in an Arg–Gln amino acid substitution at position 3611. (**B**) A sample with a heterozygous A–G mutation at the same position in the same amplicon. This plot represents median signal intensities of all overlapped probes for each base, relative to signal intensities for control samples.

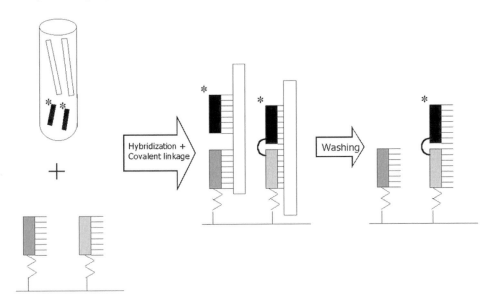

Fig. 4. Format 3 SBH. Arrays of support-bound probes are exposed to target DNAs and a labeled probe in the presence of a DNA coupling agent. Ligation of bound and labeled probes occurs only when the two probes hybridize to the target DNA at precisely adjacent positions. This ligation event results in a covalently linked two-probe complex that identifies a complementary sequence within the target DNA that is the sum of the lengths of the two individual probes.

cess was demonstrated in a blind test performed at Hyseq, funded and organized by the National Institute of Standards and Technology (NIST).

There are important advantages to the Format 3 SBH method. For example, using only two small sets of 1024 5-mer probes, researchers can determine more than one million potential 5-mer + 5-mer sequences, which is far less costly and time-consuming than testing the complete set of one million 10-mer probes directly. In addition, the ligation requires two hybridization events, thereby improving discrimination (**Fig. 5**). Finally, covalent ligation of the labeled probe to the support permits more rigorous wash conditions, which results in better signal strength relative to background. The small universal probe arrays used in Format 3 SBH are ideally suited for large-scale production by printing processes using premade probes (**Fig. 6**). A HyChip™ product using 5-mer arrays is the subject of a collaborative venture between Hyseq, Inc. and Applied Biosystems. **Figure 7** shows how the HyChip™ is used to detect sequence variations (i.e., mutations and polymorphisms) in a target DNA.

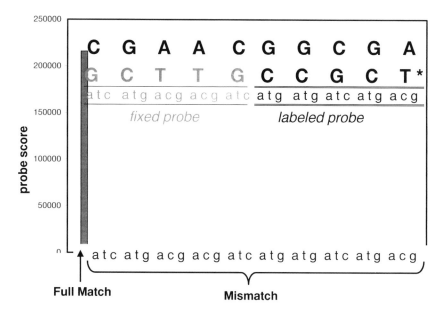

Fig. 5. Effective single mismatch discrimination. The signals of the full-match 10-mer GCTTG-CCGCT and 30 single mismatch 10-mers (3 for each of the 10 bases) are plotted in 5'–3' position order starting with the fixed probe, with a, t, c, g base order at each position. The labeled 5-mer portions of the 10-mers and its corresponding mismatches are in color. The full-match signal is >20-fold stronger than any of the 30 single mismatch signals.

4. Conclusion

Knowledge of low-frequency polymorphisms or mutations (0.1–3%), including single nucleotide substitutions, insertions, and deletions, is becoming critical to the understanding of the genetics of complex phenotypes, and to the implementation of comprehensive DNA diagnostics. To satisfy this need, hybridization of labeled probes to arrays of DNA samples (Format 1 SBH), or in combination with arrays of attached probes (Format 3 SBH), can be used to quickly and accurately sequence any selected gene from many individual DNA samples.

Acknowledgments

We wish to thank Megan Armor, Alicia Deng, Steve Huang, Helena Perazich, Julia Yeh, Tam Yen, and Ping Zhou for providing excellent experimental help; Deane Little for technical writing; Elizabeth Garnett for data presentation and graphical support; and the NIST for providing p53 mutant samples.

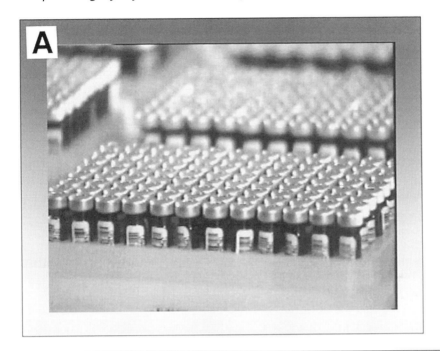

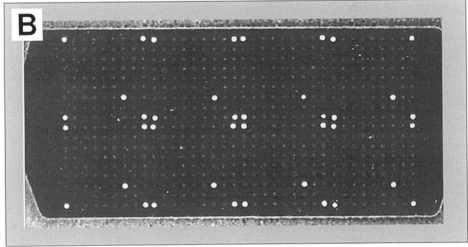

Fig. 6. Format 3 arrays prepared from premade probes. (**A**) Racks of tubes with presynthesized 5-mer probes. (**B**) Quality control of a probe array. Format 3 probe arrays are read for baseline intensities after oligonucleotides have been attached to the solid support. Both positive and negative control spots are included in such arrays, as well as replicate probes. Corners of each 9 × 9 subarray are marked by fluorescent spots. Four spots containing no probe and one orientation dot with strong fluorescence are visible in each subarray.

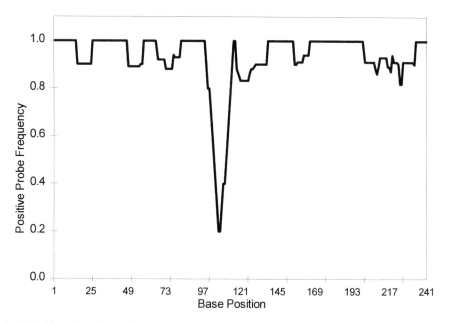

Fig. 7. Mutation detection in p53 DNA targets using Format 3 SBH. As in Format 1 SBH (*see* **Fig. 3A,B**), DNA mutations are detected by a sudden drop in positive probe frequency at a particular base position. When a mutation occurs, probes known to bind to the wild-type gene will not bind the mutant DNA at that region. The analyzed sample is an M13 clone from a NIST p53 reference mutant panel, having a CGG to TGG mutation in the codon for amino acid 248 in exon 7.

References

1. Doty, P., Marmur, J., Eigen, J., and Schildkraut, C. E. (1960) Strand separation and specific recombination in deoxyribonucleic acids: physical chemical studies. *Proc. Natl. Acad. Sci. USA* **46,** 461–466.
2. Southern, E. M. (1975) Detection of specific sequences among DNA fragments separated by gel electrophoresis. *J. Mol. Biol.* **98,** 503–517.
3. Wallace, R. B., Shaffer, J., Murphy, R. E., Bonner, J., Hirose, T., and Itakura, K. (1979) Hybridization of synthetic oligodeoxyribonucleotides to X 174 DNA. *Nucleic Acids Res.* **6,** 3543–3557.
4. Saiki, R., Walsh, P. S., Levenson, C. H., and Ehrlich, H. A. (1989) Genetic analysis of amplified DNA with immobilized sequence-specific oligonucleotide probes. *Proc. Natl. Acad. Sci. USA* **86,** 6230–6234.
5. Breslauer, K. J., Frank, R., Blocker, H., and Marky, L. A. (1986) Predicting DNA duplex stability from the base sequence. *Proc. Natl. Acad. Sci. USA* **83,** 3746–3750.
6. Drmanac, R. and Crkvenjakov, R. (1987) Method of sequencing of genomes by hybridization with oligonucleotide probes. Yugoslav patent application. (issued as US patent no. 5,202,231 [1993]).

7. Drmanac, R., Labat, I. Brukner, I., and Crkvenjakov, R. (1989) Sequencing of megabase plus DNA by hybridization: theory of the method. *Genomics* **4,** 114–128.
8. Bains, W. and Smith G. C. (1988) A novel method for nucleic acid sequencing. *J. Theor. Biol.* **135,** 303–307.
9. Southern, E. (1988) Analyzing polynucleotide sequences. International Patent Application PCT/GB 89/00460.
10. Lysov, Y. P., Florentiev, V. L., Khorlyn, A. A., Khrapko, K. R., Shick, V. V., and Mirzabekov, A. D. (1988) Determination of the nucleotide sequence of DNA using hybridization with oligonucleotides: A new method. *Dokl. Akad. Nauk. SSSR* **303,** 1508–1511.
11. Drmanac, R., Drmanac, S., Stresoska, Z., Paunesku, T., Labat, I., Zeremski, M., Snoddy, V., Funkhouser, W. K., Koop, B., Hood, L., and Crkvenjakov, R. (1993) DNA sequence determination by hybridization: A strategy for efficient large-scale sequencing. *Science* **260,** 1649–1652.
12. Drmanac, R., Stresoska, Z., Labat, I., Drmanac, S., and Crkvenjakov, R. (1990) Reliable hybridization of oligonucleotides as short as six nucleotides. *DNA Cell Biol.* **9,** 527–534.
13. Drmanac, R. and Drmanac, S. (1999) cDNA screening by array hybridization. *Meth. Enzymol.* **303,** 165–178.
14. Drmanac, S., Kita, D., Labat, I., Hauser, B., Burczak, J., and Drmanac, R. (1998) Accurate sequencing by hybridization for DNA diagnostics and individual genomics. *Nat. Biotechnol.* **16,** 54–58.
15. Drmanac, R. (1994) PCT/US94/10945 patent application.

4

Ethical Ramifications of Genetic Analysis Using DNA Arrays

Wayne W. Grody

1. Introduction

1.1. Our Eugenics Legacy

Since its earliest days, the history of human genetics has been checkered with actual, perceived, and potential abuses in the application of its scientific concepts to research or clinical endeavors. Aside from obvious cases of scientific fraud and continuing controversies over natural selection and biological determinism, a long and varied history of eugenics movements grew out of the (re)discovery of Mendel's experiments at the beginning of the twentieth century. The term *eugenics* was first coined by Francis Galton in 1883 and defined as the science of improving the gene pool of the human species through selective breeding. The concept was soon extended well beyond its obvious and accepted precedent of animal husbandry to encompass social as well as physical traits, as Mendelian inheritance came to be viewed as the fundamental determinant of low intelligence, mental illness, substance abuse, physical handicaps, poverty, promiscuity, prostitution, and criminality. Eugenics thus provided a logical extension to the notion that such undesirable traits could be weeded out of the population through biological means rather than traditional social welfare policy, by enactment of restricted marriage laws, mandatory sterilization, euthanasia, or outright genocide. Of course, everyone knows that such policies reached their zenith (or nadir) in Nazi Germany, where the scope of supposedly genetic social traits to be eliminated through these means expanded to include race and ethnicity. Unfortunately, the horrific scale of the Nazi offenses has made it all too easy to forget that influential eugenics movements existed in many countries of the world in the early decades of the twentieth

From: *Methods in Molecular Biology, vol. 170: DNA Arrays: Methods and Protocols*
Edited by: J. B. Rampal © Humana Press Inc., Totowa, NJ

century, including the United States, which enacted restrictive marriage and immigration laws, performed countless mandatory sterilizations of criminals and the mentally retarded, and showcased racially and socially ideal procreation through "Fitter Family" competitions at state fairs *(1)*. And lest we get too complacent in relegating such behaviors to the quaint early years of the century, the concept of "racial hygiene" invoked by the Nazis has resurfaced quite literally in the new term *ethnic cleansing* being dispatched in Eastern Europe at the time of this writing.

While such gross abuses are now considered taboo in Western countries, the advent of new genetic technologies and the robust economics to pursue them has raised the specter of a new, more subtle type of eugenics practice based on specific gene selection and replacement at the molecular level. Moreover, because these technologies are expensive and resources are not always equitably distributed, discriminatory eugenic practices may be effected unwittingly, simply by virtue of unequal access.

1.2. The New Diagnostic Molecular Genetics

Given that modern debate over eugenics issues is technology driven, it is worth considering how present molecular genetic techniques have evolved to the point where such issues come to the fore. The new genetics, a long-term outgrowth of the elucidation of the double-helical structure of DNA by Watson and Crick almost 50 yr ago, comprises two arms: gene-level diagnostics and gene-level therapeutics. These two efforts have developed at different rates and have achieved different degrees of success, with gene therapy a more recent and technically more problematic endeavor. In one sense the two go hand in hand, because gene replacement therapy cannot be considered until a patient's precise molecular defect has been determined. But for the purposes of this chapter, in a book on oligonucleotide arrays applied to gene structure, expression, and discovery, the first arm, gene diagnostics, is the more relevant.

The field of clinical molecular diagnostics has been growing rapidly since the early 1980s, to the point where it is now a recognized and respected subdiscipline of laboratory medicine, replete with its own dedicated journals, subspecialty board certifications, laboratory accreditation and quality assurance programs, and a specialized scientific organization, the Association for Molecular Pathology. Hundreds of individual DNA-based tests are available in both academic and commercial laboratories, spanning applications in infectious disease, cancer, genetic disease, and DNA fingerprinting. Of these, it is molecular testing for genetic disease that has provoked the most discussion over ethical questions, in part because of the earlier history of eugenics and also because, by its very nature, a DNA test for a genetic disease involves detection of inherited alterations of the germline, which comprises a person's funda-mental genetic makeup. Concerns over invasion of "genetic privacy," stigmatization, and discrimination naturally follow. By contrast, the use of molecular techniques to detect the foreign genomic sequence of an invading microorganism, or even the deranged genetic makeup of a malignant tumor cell, does not raise such concerns in quite the same way.

Eugenics practices, by definition, apply to large populations, and until recently, the standard techniques available to diagnostic molecular genetics laboratories were too cumbersome and expensive for widespread application to be feasible. The predominant technique for many years, the Southern blot, is extremely labor-intensive, not amenable to specimen batching or high throughput, and typically reliant on radioisotopic signal detection, which adds to its expense and awkwardness. The Southern blot is a means to assess the gross size and structure of a gene, but the same limitations apply to the techniques previously available for fine-structure analysis of large, complex genes exhibiting mutational heterogeneity. Both DNA sequencing and the various mutation scanning techniques, such as single-strand conformation polymorphism analysis, denaturing gradient gel electrophoresis, and the protein truncation test, are technically difficult, involved, multistep processes; the scanning techniques, in addition, are notoriously inefficient and will miss a fair proportion of mutations. Automation of DNA sequencing has certainly reduced its labor requirements and improved throughput, but the per-test expense is still high, as evidenced by current charges for complete sequencing of the *BRCA1* and *BRCA2* genes in women at risk for familial breast/ovarian cancer. No test costing thousands of dollars, even if automated, could be applied to population screening.

Two developments in molecular genetic technology, one of which is the subject of this book, now promise to bring these tests to the masses. The first was the advent of the polymerase chain reaction (PCR), which vastly increased the sensitivity and specificity of molecular genetic tests while also markedly lowering their expense and freeing them from dependence on costly and potentially hazardous radioisotopes. By reducing the bulk of the specimen workup to a single small microtube (often followed by simple dot-blot hybridization with allele-specific oligonucleotide probes), the test procedures became easier to handle. With the subsequent appearance of automated thermocyclers able to process many such reaction vessels in parallel or to accept samples loaded into 96-well microtiter plates, throughput was further enhanced and molecular genetic testing of large numbers of people became feasible. Still, the analysis was limited to detection of one or a few mutations in each sample at a time, so the overall yield for analysis of large, heterogeneous genes such as *BRCA1* or *CFTR* remained suboptimal. This remaining deficiency now looks to be circumvented by the second key development, the oligonucleotide array or DNA

Table 1
Molecular Classification of Genetic Disorders

1. Diseases for which both the gene and mutation are known
2. Diseases for which the gene is known, but not the mutation
3. Diseases for which neither the gene nor the mutation is known
4. Diseases caused by more than one gene (polygenic)

chip. Just as PCR enables molecular genetic analysis of many patient specimens in parallel, microarray technology will enable analysis of countless mutations and genes in parallel on each specimen. The potential thus exists, by combining both technologies, for complete gene scanning and detection of all possible mutations in a large-scale screening program.

1.3. Application of Array Technology to Molecular Genetic Testing

In deciding which technique to employ for molecular genetic testing, laboratories must consider what is known about the particular disease gene in question. Using this approach, every disorder can be assigned to one of four categories, as listed in **Table 1**. The simplest tests are those for disorders in which the gene is known as well as the mutation being searched for, because every patient with that disorder has the same mutation. Examples include sickle cell anemia, achondroplastic dwarfism, and thrombophilia owing to clotting factor V-Leiden mutation. These diseases are amenable to mutation detection by simple PCR-based tests that hone in on the precise nucleotide mutation site, either by allele-specific oligonucleotide probe hybridization or by restriction endonuclease digestion of the amplicons in cases in which the mutation is known to destroy or create a restriction enzyme recognition sequence. Much more difficult are disorders for which the gene is known but not the mutation, because different patients with the same disease may have any number of different mutations. Most of the newer disease genes being elucidated through the Human Genome Project fall within this category, a prime example being the *CFTR* gene associated with cystic fibrosis (CF), in which more than 800 mutations have been reported to date. Still more difficult are disorders for which the causative gene has not yet been identified. If it has at least been mapped to a particular chromosome, prenatal diagnosis and carrier detection can be offered in certain families using linkage analysis, tracking polymorphic DNA markers that are located close to the unknown disease gene on the same chromosome. Finally, there are the polygenic/multifactorial disorders, such as atherosclerosis and diabetes, believed to be caused by several genes interacting with one another and with exogenous environmental factors. At least until more of the relevant genes are discovered and their interactions are understood, this class

Table 2
Clinical Applications of Molecular Genetic Testing

1. Clinical diagnosis/differential diagnosis
2. Carrier screening
3. Prenatal diagnosis
4. Newborn screening
5. Presymptomatic diagnosis/predisposition screening

of disorders will remain beyond the capabilities of our present molecular diagnostic technology.

The development of oligonucleotide array technology holds great promise for surmounting the technical obstacles to cost-effective genetic analysis of the second disease category in **Table 1** and perhaps eventually the fourth as well. By hybridization of patient DNA to a large number of oligonucleotide probes simultaneously, it should be possible to detect, e.g., all 800 known mutations of the *CFTR* gene, and many others yet to be discovered as well. Similarly, the same chip could be used to detect mutations in many other genes in the same specimen at the same time. Some of these mutations may be associated with disorders for which the individual did not know he or she was at risk or was even going to be tested. This power to obtain a vast amount of genetic information on an individual in a matter of minutes is at the heart of any discussion of the ethical issues surrounding genetic testing using DNA arrays.

Even with routine technology, the psychosocial impact of molecular genetic testing varies according to the clinical intent of the testing. **Table 2** lists the clinical applications of molecular genetic testing. The first is also the most straightforward: laboratory confirmation of a clinical diagnostic impression in a symptomatic patient is the basic activity of all of laboratory medicine. In the genetic disease area, however, even this noncontroversial activity raises unique ethical issues, because the detection of a germline mutation has implications that extend beyond the immediate patient being tested to include all of that person's blood relatives (not all of whom may wish to know this information). By contrast, carrier screening for recessive mutations has no symptomatic significance even for those testing positive, but does have profound implications for their reproductive decision making. Thanks to the new technologies already discussed, such screening can now be conducted across large populations, placing unprecedented demand on existing genetic counseling resources.

Prenatal diagnosis has been aided greatly by PCR, the sensitivity of which allows for molecular testing on leftover amniocytes obtained for other purposes, or on tiny chorionic villus samples, or even on fetal cells circulating in the mother's blood. A positive result, however, usually leads to difficult decisions about termination of the pregnancy, with all the attendant ethical and legal controversies surrounding access to abortion.

Until recently, newborn screening for treatable metabolic disorders such as phenylketonuria and galactosemia has been performed by biochemical analysis on dried blood spots, but PCR-based testing is also possible on such samples, and some states have begun to institute molecular genetic testing as a complement or backup to the biochemical assays. The availability of array technology would allow for screening of newborns for many more diseases than the small numbers now targeted (three or four in most states). Because some form of newborn screening is mandatory in all 50 states, issues of coercion and autonomy will naturally be raised as additional diseases, each with its own potential for stigmatization, are incorporated into the screening program. It should also be kept in mind that DNA-based newborn screening has the potential to identify not only affected infants but carriers as well, potentially stigmatizing unnecessarily both the children themselves and their parents.

Perhaps the most daunting clinical application is presymptomatic diagnosis of later-onset autosomal dominant disorders (sometimes called predisposition testing in the case of mutations with <100% penetrance). Prominent examples include Huntington disease, for which no treatment can be offered to those testing positive, and familial breast/ovarian cancer (*BRCA1* and *BRCA2* genes), in which both the surveillance and treatment interventions remain controversial. DNA arrays could potentially include mutation probes for a large panel of late-onset disorders, dramatically complicating an already thorny genetic and social counseling problem.

2. General Ethical Principles Guiding Genetic Testing and Research

Most of the ethical principles to which we now adhere in the treatment of both patients and medical research subjects grew, ironically, out of the most shameful episodes in the history of the field (*see* **Subheading 1.**) As a direct result of the revelation of Nazi abuses at the end of World War II, the Nuremberg Code was drafted by international consensus *(2)*, restoring individual autonomy to potential subjects of medical research or interventions. Further elaboration of the desired ethical principles of biomedical research (the triad of respect for persons, beneficence, and justice) was formalized in the United States in the famous Belmont report of 1979 *(3)*. These and other professional guidelines have led to our present state of respect for the individual patient or research subject and to the notions of voluntary participation, informed consent, right to privacy and confidentiality, and minimization of risk of direct harm or later discrimination. Adherence to these standards is monitored closely by federal agencies, institutional review boards, and hospital ethics boards.

Despite these safeguards, the newfound power of molecular genetic analysis has suggested to some that additional regulations may be needed. As alluded to previously, molecular genetic testing represents more than a single discrete research or clinical intervention, because the information obtained may predict future events or affect the lives of others besides the person consenting to be tested. Moreover, given the physical stability of DNA and the ability of PCR to work on minuscule amounts of even degraded specimens, such testing can now be performed without even the subject's knowledge, let alone consent. A specimen obtained previously, even many years earlier, for some relatively innocuous clinical purpose or unrelated research project, could now be used to conduct the most potentially devastating sort of analysis, such as predictive testing for cancer or dementia. Even when performed with appropriate consent, predictive genetic testing represents a radical departure from the traditional paradigm of laboratory medicine, in that disease (or high risk of disease) may now be diagnosed years or decades before the appearance of the first sign or symptom—long before the first choreic movement of Huntington disease or the transformation of the first malignant cell in hereditary breast cancer. Such predictive knowledge forces us to reexamine, in an almost philosophical way, our traditional definition of disease and, for insurance purposes, the definition of a preexisting condition. It is the realization of this impending paradigm shift that is behind the many pieces of federal and state legislation currently being introduced to ensure genetic privacy, nondiscrimination, and confidentiality. It is also the rationale behind the creation of a special branch of the US Human Genome Project, the Ethical, Legal and Social Implications (ELSI) program, to monitor and fund studies to address these issues *(4)*. How much more complex will these efforts become when oligonucleotide arrays expand our diagnostic and predictive power by many orders of magnitude?

3. Ethical Concerns Raised by Clinical Use of Oligonucleotide Arrays

The introduction of oligonucleotide arrays into both the clinical and research settings raises unprecedented questions about the scope and limits of genetic testing in patients and subjects. This section considers the most important of these issues, many of which have already been introduced, as they pertain to large array analysis.

3.1. Appropriateness of Test Ordering and Reporting

Modern molecular genetic tests, especially those predicting risk of future disease, present a formidable challenge to physicians who must attempt to keep up with almost weekly breakthroughs in the field, judiciously triage patients who would be suitable candidates for the tests, order them appropriately and send the specimens to a qualified laboratory, and interpret and report the subtle complexities of the test results back to the patient. Given that most of the scientific advances in this field occurred well after the graduation of most practicing physicians from medical school, and that genetics is not particularly well taught in medical schools even now *(5)*, there is considerable trepidation that general physicians are not up to the task.

Indeed, a recent survey of predictive molecular genetic testing for familial adenomatous polyposis revealed that large proportions of patients were given incorrect information about the results of the test *(6)*. As this sort of testing continues to proliferate, referral of all such cases to a medical genetics clinic for proper interpretation becomes less of a practical option, because there are not enough genetic counselors in the all of the United States to handle the anticipated case load of even a single large program, such as nationwide carrier screening for cystic fibrosis mutations *(7)*. Predictive testing for the *BRCA1* and *BRCA2* gene mutations associated with familial breast/ovarian cancer commonly requires 1 to 2 h of pre- and post-test genetic counseling to explain all the complexities of reduced penetrance of the mutations, the inability to detect all possible mutations in the genes, the possible involvement of other genes not discovered yet, the uncertain options for clinical interventions in those testing either positive or negative, and so on *(8)*. Moreover, as DNA arrays become larger and more pervasive in the testing milieu, counseling demands will increase exponentially, because each individual test request will generate a myriad of genotypic results that need to be interpreted for the patient.

These concerns were behind the creation of the NIH-DOE Task Force on Genetic Testing by the ELSI branch of the Human Genome Project. This body deliberated for two years to develop guidelines for ensuring the quality and appropriate use of genetic tests, especially those of a predictive nature *(9)*. The Task Force's published final report *(10)* contains many recommendations for validation of new genetic tests, quality assurance procedures, improved human genetics education of the public and the profession, confidentiality of test results, and informed consent. These recommendations have generally been well received by the medical genetics community, but, for the most part, they were generated within the context of a single disease test performed at a time. Therefore, it is unclear how well these recommendations will translate to the testing of many genes and mutations in parallel using microarrays. Will every genotypic analyte on the DNA chip represent an equally appropriate use of genetic testing for the patient in question? How will care providers be able to interpret and counsel so many disparate test results within a practical time frame? How can all these gene tests be validated and quality controlled to the same degree, or at least to an acceptable level? Note that the test validation recommended by the task force is of two types: analytic validation (does the test detect the mutation it is designed to analyze?) and clinical validation (how well does detection of the mutation actually predict disease?). The latter is much more difficult, sometimes requiring decades of intense study and clinical follow-up. Clearly our ability to add more and more mutation probes to an array will rapidly outstrip our ability to clinically validate each of them.

3.2. Automization

The Task Force on Genetic Testing *(10)*, the American Society of Clinical Oncology *(11)*, and numerous other bodies have strongly endorsed the essential importance of pre- and post-test counseling for complex gene tests such as those for familial cancer predispositions. The few laboratories offering such tests at present require documentation of counseling before they will run the analysis. There is no question that array technology will make the testing for these complex genes, with their hundreds of possible mutations, much easier, and probably more comprehensive, than the brute-force DNA sequencing approaches now used. The same will be true for large-scale carrier screening of genes with many possible mutations, such as that for CF; in this setting, too, adequate counseling has been deemed essential *(7)*. There is legitimate concern, however, that array technology will make such testing too easy, that by automating the entire analytic process the testing could too quickly move into the high-volume clinical laboratory setting, breaking the traditional connection between specialized molecular diagnostic laboratories and their associated genetic counselors. As the field evolves in this direction, every effort must be made to educate the clinical chemistry community about the need for vigilant gatekeeping to ensure appropriate test ordering and applicable informed consent, as well as adequate test interpretation and post-test counseling.

4. Informed Consent and Patient Autonomy

As a result of the Nuremberg Code and the promulgation of individual autonomy as paramount, the concept of voluntary participation and informed consent for both medical and research procedures has been widely accepted. All patients admitted to the hospital or visiting the outpatient clinic sign a consent to cover subsequent diagnostic and therapeutic procedures. Given the unique nature of genetic testing, many institutions are now modifying their admission consent forms to address genetic issues specifically. There has been some disagreement in genetic circles over which types of tests require separate informed consent above and beyond the blanket consent administered at the time of admission. Some feel that any and all genetic tests require consent, whereas others, including this author, believe that separate consent is not necessary for diagnostic tests but only for predictive ones *(12)*; the Task Force on Genetic Testing has also addressed this issue *(10)*. The introduction of DNA arrays to the mix will complicate this debate even further, because a single array may well contain probes that are both diagnostic and predictive. For example, an array designed to assess various oncogene mutations and other somatic changes in tumor DNA could just as easily pick up a germline change in a tumor suppressor gene at the same time, changing a diagnostic/prognostic test to a predictive one. At least one tumor suppressor gene, *p53*, may exhibit either somatic or inherited mutations. Of course, one could simply choose not to report DNA findings considered extraneous to the main purpose of the test, but failure to divulge test results brings up other ethical quandaries. Whatever the decision, the replacement of single-mutation and single-gene tests with complex arrays will require much more elaborate and detailed informed consent procedures.

4.1. Genotype-Phenotype Correlation and Predictive Value

The problem of reduced penetrance of many dominant mutations, such as those in the familial breast/ovarian and colon cancer genes, has been alluded to, making clinical validation and counseling on test results much more difficult. Even some recessive mutations have this problem; the genotype-phenotype correlation for CF mutations, for example, is notoriously poor, so that it is difficult to predict for parents the likely severity of their child's disease. This is one of the major factors behind the objection, in some quarters, to institution of nationwide CF carrier screening, despite the relatively high carrier frequency in the general population *(7)*.

For those who favor CF screening, a "CF chip" is eagerly awaited as a solution to the present inability to detect more than a small fraction of the hundreds of possible mutations in the *CFTR* gene. But a truly comprehensive DNA array brings with it another problem: when it becomes possible to detect every potential nucleotide change in a gene, the test will reveal polymorphisms as well as mutations. If the DNA alteration detected has not been reported before in the context of the disease phenotype, it may be difficult or impossible to decide whether it represents a pathologic mutation or merely a benign polymorphism. Obviously, a nonsense or frameshift mutation, especially near the 5' end of the gene, can be presumed to be deleterious, but what about a single nucleotide substitution? Even if it encodes a nonconservative substitution of a different chemical class of amino acid, it could be located in a nonessential domain of the protein and therefore harmless. This is the reason many *BRCA* sequencing tests produce a result reported as "DNA change of uncertain significance" *(13)*. When arrays make it possible to detect many more single nucleotide changes all over the genome, the reporting criteria will become even more complicated. The problem is one of more than academic interest, because the patient is left to labor under this uncertain result for the rest of his or her life, not knowing if it increases his or her risk of disease or not.

4.2. Insurance and Employment Discrimination

Documentation on the risk of losing access to health, life, or disability insurance, or employment, based on genetic testing results, was first brought to light by Billings et al. *(14)*. Interestingly, not all their cases involved presymptomatic tests for adult-onset dominant disorders; some were carrier screens for recessive mutations, in which those testing positive will never be symptomatic. Although many commentators feel that life insurance is a privilege rather than a right, and that life insurers should have some protection against "adverse selection," in which the proposed insured possesses predictive genetic knowledge that the company does not, a general consensus has arisen that it is improper to restrict access to health insurance based on genetic test information. This has led to pleas for insurance reform at the national and state levels *(15)*, and, indeed, a number of states have already passed such protective legislation.

These initiatives are certainly valid, although some observers *(16)*, including this author, have felt that the actual magnitude of the threat has been somewhat exaggerated; most geneticists would be hard pressed to think of a single case of insurance discrimination in their own experience, despite the many thousands of patients who have undergone such testing. Moreover, the insurers themselves deny that they have any interest in pursuing molecular genetic testing as a means of restricting or stratifying individuals for health insurance, stating that they are more worried about the millions of people with high cholesterol than the relatively small number with single-gene disorders (strictly speaking, though, cholesterol measurement may also be considered a genetic test, albeit a nonmolecular one). But a skeptic might retort that the only reason insurance companies are not screening for discrete mutations at present is that it is too expensive to test for all the relevant genes and mutations associated with even the more prevalent disorders. Would their attitude change when oligonucleotide arrays make such screening feasible? Can we envision a day when DNA chip testing will be a part of routine newborn screening, and insurance companies will ascertain on the day of birth which future diseases that person will fall prey to? Because we are all estimated to carry, on average, six deleterious mutations, how will the industry decide who is worth insuring? In actuality, it is employment rather than the insurance setting that may pose more of a threat. When gene expression arrays are able to identify subtle pharmacogenetic differences among individuals in their ability to metabolize drugs and toxins, or to predict personality traits or future behavioral disorders, will companies use such testing to exclude potential employees from certain occupations?

4.3. Testing of Children and Newborns

At least one of the aforementioned hypothetical scenarios would be met with significant objection from the genetics community if it were to be instituted: the random testing of newborns with DNA arrays. Even for single-gene defects for which a child is at risk based on family history, there is a strong consensus that predictive or carrier testing should not be performed until the age of consent, unless there is an accepted preventative medical intervention that would need to be initiated during childhood to be effective. Long an unwritten rule in the genetics community, this notion has now been codified in the report of the Task Force on Genetic Testing *(10)*. The argument would be even stronger with regard to array screening, in which any number of gene defects would be tested blindly, without regard to family history. This consensus derives in part from concern over possible insurance discrimination, but even more so from fear of social stigmatization of a child at an age when he or she cannot understand the meaning of the genetic information nor has any practical need for it.

4.4. Genetic Privacy and Confidentiality

Regardless of how or why a molecular genetic test is performed, whether in the clinical or research setting, maintenance of individual privacy through adequate security and strict confidentiality is essential. For some highly charged predictive tests, results may be given only to the patient and not placed in the medical record or conveyed to the referring physician. Laboratories must also use caution in the delivery of written reports, avoiding nonsecure computer systems or faxes *(12)*. As hospitals incorporate paperless (i.e., electronic) medical records, the security issues become more acute. Recently some concern even has been raised over possible violations of confidentiality by the publication of pedigrees in journal articles, in which a unique pattern might be seen and recognized by other family members, thereby unwittingly revealing sensitive phenotypic or genetic testing information. New editorial procedures have been proposed for the journals, possibly including alteration of certain aspects of the pedigrees to disguise them *(17,18)*.

Confidentiality concerns will only increase as automated large-scale arrays depersonalize the testing process and mandate data dumping and storage in huge institutional information systems. It is likely that the genetic information produced by such tests will be so voluminous and complex that no single written report or one-on-one counseling session could adequately record and convey the content. Patients and their physicians may instead have to refer back repeatedly to the computerized database—the modern and much more complex version of the medical alert bracelet. But where should such a database be stored, and who should have access to it? Most likely the necessary precautions will dovetail with pending regulations designed to ensure privacy of the computerized medical record.

4.5. Population and Ethnicity Issues

Privacy and stigmatization concerns apply not only to individuals but also to populations. Medical genetics is probably the least "politically correct" of all specialties, in that ethnic differences in allele frequencies are a fact of life and cannot be brushed aside. As more ethnic-specific mutations are identified, risks of insurance and other discrimination have become a concern not only for the individual being tested, but also for every member of that person's ethnic group. In this way, Ashkenazi Jewish women, for example, have become worried that they may be redlined for health insurance because of the discovery of a high-carrier frequency for certain mutations in the *BRCA1* and *BRCA2* genes in this ethnic group *(19)*. Some authors have recently proposed that, prior to initiating a genetic research project in a particular ethnic group, a dialogue should be established with the community and a sort of "group consent" (in addition to the individual participant consents) should be obtained *(20)*. But how should the "group" be defined in our ethnically mixed population, and who should be deemed qualified to speak on behalf of it?

The widespread adoption of array technology for genetic testing will reveal an ever-increasing constellation of ethnic mutations and polymorphisms. Furthermore, with the huge capacity of DNA chips, a single array can easily encompass the mutations of any number of ethnic groups. This capacity will likely drive the manner in which population screening programs are administered—the availability of a "CF chip," for example, would make panethnic CF carrier screening more feasible and just as easy (or even easier) as specific targeting of Caucasians or other high-risk groups *(7)*. But what sort of consent and privacy provisions should be maintained? And what is the mechanism for one ethnic group, or a few individuals from that group, to opt out? These questions, as well as the additional cost such procedures would impart, may well make the group consent model untenable, except in a few small and well-defined ethnic populations *(16,21)*.

4.6. Prenatal Diagnosis and Abortion

Prenatal diagnosis always carries with it the possible endpoint of abortion of an affected fetus. The same endpoint is also key in large-scale population screening programs for recessive mutations, in which couples at risk are identified and offered prenatal diagnosis. Regardless of one's moral and political views about access to abortion, the procedure is a serious step that few would take lightly, whatever the circumstances. It almost goes without saying that, in order to contemplate taking such a step, the disorder being tested for should be a severe one—if not incompatible with life, then incompatible with a normal or tolerable life. The controversies over nationwide CF carrier screening have already been mentioned in this regard. DNA arrays now open up the possibility of screening couples and fetuses for all sorts of traits beyond the most severe disease genotypes. They could theoretically give couples new eugenic options for avoiding offspring with mild quantitative traits that many would not consider serious enough to justify termination. The availability of preimplantation diagnosis, with termination of affected embryos in vitro, allows concerned couples to be one step removed from the onus of abortion, but the procedure is expensive and not always successful (22). Will we need to develop a list of diseases whose phenotype is deemed serious enough to qualify for inclusion on the prenatal DNA chip? Who will decide which diseases are to be included on this list, given that severity is perceived differently by different people and changes over time as treatments evolve?

4.7. Access to Stored Samples

Because PCR-based genetic analysis can be performed on stored, even fixed, tissue or blood samples, concern has arisen, first articulated in a series of workshops sponsored by the ELSI program, about the potential for violation of privacy by performing genetic testing and research on such samples long after they were donated, without the donor's knowledge or consent. The initial proposals emanating from those workshops recommended a quite detailed informed consent process, by tracking down the original donors or their next of kin, prior to studying any stored materials that were not anonymous or anonymized (23). Others felt that these measures tipped the balance too far toward the side of individual autonomy at the expense of the rights of society as a whole to benefit from the tremendous research resource that such collected material represents, and alternative proposals were put forward (24–28). The use of stored materials from individuals of known genotype or phenotype is essential to the continuing development and validation of genetic tests, in addition to increasing our understanding of the molecular basis of genetic, neoplastic, and infectious disease. Some of the early proposals required specificity of consent down to the particular genes to be studied, which would clearly exclude any research done with large DNA arrays. This would be a great loss, because one of the most promising applications of arrays is in the study of thousands of genes in parallel, especially toward the goal of generating individualized tumor profiles and pharmacogenetic response patterns.

5. Economics and Reimbursement

Reimbursement for clinical molecular genetic testing has been a touch-and-go affair for the laboratories involved. For the most part, as stated at the outset, these tests are expensive to perform, are often conducted on healthy individuals, and are considered esoteric, with few if any kits licensed by the Food and Drug Administration (FDA). Thus, third-party payers have plenty of excuses to avoid reimbursement. Like any new technology, oligonucleotide arrays will be expensive when first introduced, and the reimbursement levels may not be adequate for laboratories to recover their costs. Discriminate access to testing, based on socioeconomic level, will be a recurring ethical consideration, at least in the beginning.

Some progress in reimbursement for molecular diagnostics has occurred recently with the acceptance of new billing codes that more accurately represent the scope and effort of current techniques *(29)*. The FDA has also recently attempted to come to terms with the predominance of nonlicensed, "home brew" tests in the field by making allowances for the use of "analyte-specific reagents" (such as DNA probes and PCR primers) as components of such tests *(30)*. A current concern arises from the increasing number of disease genes and mutations that are subject to patents and exclusive licenses; whether through monopolistic test delivery or mandatory royalties, this situation can only serve to increase the cost of testing, perhaps prohibitively. Taken to its extreme, the patent crisis could kill clinical oligonucleotide array technology before it even gets started: What would be the per-test charge if patent royalties had to be paid on each of 100,000 sequences present on the DNA chip?

6. Conclusion

There is little question that oligonucleotide arrays represent the next big step in molecular genetic testing and research, as monumental in their own way as the introduction of PCR was in the mid-1980s. As was true for PCR, the advent of a new technology of such great power brings with it a panoply of ethical questions. We should not be intimidated by such questions but, rather, should use them to guide us in formulating approaches to ensure that the technology remains available to benefit the greatest number of patients and to enhance research on the molecular basis of human disease. As in most areas of scientific endeavor, worries about discrimination and other potential abuses of the technology should be directed not at the science itself but at the societal setting of its application. As long as appropriate but not overly burdensome protections for patients and research subjects are in place, and economic mechanisms exist to ensure equal access, oligonucleotide array technology should take its rightful place as the predominant mode of molecular genetic testing, screening, and monitoring well into the 21st century.

References

1. Kevles, D. J. and Hood, L. (ed.) (1992) The *Code of Codes: Scientific and Social Issues in the Human Genome Project,* Harvard University Press, Cambridge, MA.
2. Nuremberg Code. (1949) *Trials of War Criminals Before the Nuremberg Military Tribunals Under Control Council Law No. 10,* vol. 2, U.S. Government Printing Office, Washington, DC.
3. National Commission for the Protection of Human Subjects of Biomedical and Behavioral Research. (1979) *The Belmont Report: Ethical Principles and Guidelines for the Protection of Human Subjects of Research,* U.S. Department of Health, Education and Welfare, Washington, DC.
4. United States Department of Energy. (1997) Human *Genome Program Report, Part 1: Overview and Progress,* U.S. Department of Energy, Germantown, MD.
5. Grody, W. W., Kronquist, K. E., Lee, E. U., Edmond, J., and Rome, L. H. (1993) PCR-based cystic fibrosis carrier screening in a first-year medical student biochemistry laboratory. *Am. J. Hum. Genet.* **53,** 1352–1355.
6. Giardiello, F. M., Brensinger, J. D., Petersen, G. M., Luce, M. C., Hylind, L. M., Bacon, J. A., Booker, S. V., Parker, R. D., and Hamilton, S. R. (1997) The use and interpretation of commercial *APC* gene testing for familial adenomatous polyposis. *N. Engl. J. Med.* **336,** 823–827.
7. Grody, W. W. (1999) Cystic fibrosis: Molecular diagnosis, population screening, and public policy. *Arch. Pathol. Lab. Med.* **123,** 1041–1046.
8. Geller, G., Botkin, J. R., Green, M. J., Press, N., Biesecker, B. B., Wilfond, B., Grana, G., Daly, M. B., Schneider, K., and Kahn, M. J. E. (1997) Genetic testing for susceptibility to adult-onset cancer: The process and content of informed consent. *JAMA* **277,** 1467–1474.
9. Holtzman, N. A., Murphy, P. D., Watson, M. S., and Barr, P. A. (1997) Predictive genetic testing: From basic research to clinical practice. *Science* **278,** 602–605.
10. Task Force on Genetic Testing. (1998) *Promoting Safe and Effective Genetic Testing in the United States,* Johns Hopkins University Press, Baltimore, MD.
11. American Society of Clinical Oncology. (1996) Statement of the American Society of Clinical Oncology: Genetic testing for cancer susceptibility. *J. Clin. Oncol.* **14,** 81,730–81,736.
12. National Committee for Clinical Laboratory Standards. (1999) *Molecular Diagnostic Methods for Genetic Diseases: Approved Guidelines,* National Committee for Clinical Laboratory Standards, Wayne, PA.
13. Shattuck-Eidens, D., Oliphant, A., McClure, M., McBride, C., Gupte, J., Rubano, T., et al. (1997) BRCA1 sequence analysis in women at high risk for susceptibility mutations: Risk factor analysis and implications for genetic testing. *JAMA* **278,** 1242–1250.
14. Billings, P., Cohn, M., de Cuevas, M., Beckwith, J., Alper, J., and Natowicz, M. (1992) Discrimination as a consequence of genetic testing. *Am. J. Hum. Genet.* **50,** 476–82.
15. Hudson, K. L., Rothenberg, K. H., Andrews, L. B., Kahn, M. J. E., and Collins, F. S. (1995) Genetic discrimination and health insurance: An urgent need for reform. *Science* **270,** 391–393.

16. Reilly, P. R. (1998) Rethinking risks to human subjects in genetic research. *Am. J. Hum. Genet.* **63,** 682–685.
17. Botkin, J. R., McMahon, W. M., Smith, K. R., and Nash, J. E. (1998) Privacy and confidentiality in the publication of pedigrees: A survey of investigators and bio-medical journals. *JAMA* **279,** 1808–1812.
18. Byers, P. H. and Ashkenas, J. (1998) Pedigrees—Publish? or perish the thought? *Am. J. Hum. Genet.* **63,** 678–681.
19. Struewing, J. P., Abeliovich, D., Peretz, T., Avishai, N., Kaback, M. M., Collins, F. S., and Brody, L. C. (1995) The carrier frequency of the BRCA1 185delAG mutation is approximately 1 percent in Ashkenazi Jewish individuals. *Nat. Genet.* **11,** 198–200.
20. Foster, M. W., Bersten, D., and Carter, T. H. (1998) A model agreement for genetic research in socially identifiable populations. *Am. J. Hum. Genet.* **63,** 696–702.
21. Juengst, E. T. (1998) Group identity and human diversity: Keeping biology straight from culture. *Am. J. Hum. Genet.* **63,** 673–677.
22. Pembrey, M. E. (1998) In the light of preimplantation genetic diagnosis: Some ethical issues in medical genetics revisited. *Eur. J. Hum. Genet.* **6,** 4–11.
23. Clayton, E. W., Steinberg, K. K., Khoury, M. J., Thomson, E., Andrews, L., Kahn, M. J. E., Kopelman, L. M., and Weiss, J. O. (1995) Informed consent for genetic research on stored tissue samples. *JAMA* **274,** 1786–1792.
24. Knoppers, B. M. and Laberge, C. M. (1995) Research and stored tissues: Persons as sources, samples as persons? *JAMA* **274,** 1806–1807.
25. Grody, W. W. (1995) Molecular pathology, informed consent, and the paraffin block. *Diagn. Molec. Pathol.* **4,** 155–157.
26. Grizzle, W., Grody, W. W., Noll, W. W., Sobel, M. E., Stass, S. A., Trainer, T., Travers, H., Weedn, V., and Woodruff, K. (1999) Recommended policies for uses of human tissue in research, education, and quality control. *Arch. Pathol. Lab. Med.* **123,** 296–300.
27. American Society of Human Genetics. (1996) ASHG Report: Statement on informed consent for genetic research. *Am. J. Hum. Genet.* **59,** 471–474.
28. Merz, J. F., Sankar, P., Taube, S. E., and LiVolsi, V. (1997) Use of human tissues in research: Clarifying clinician and researcher roles and information flows. *J. Invest. Med.* **45,** 252–257.
29. Grody, W. W. and Watson, M. S. (1997) Those elusive molecular diagnostics CPT codes. *Diagn. Mol. Pathol.* **6,** 131–133.
30. Garrett, C. T. and Ferreira-Gonzalez, A. (1996) FDA regulation of analyte-specific reagents (ASRs): Implications for nucleic acid-based molecular testing. *Diagn. Mol. Pathol.* **5,** 151–153.

5

Photolithographic Synthesis of High-Density Oligonucleotide Arrays

Glenn H. McGall and Jacqueline A. Fidanza

1. Introduction

High-density polynucleotide probe arrays provide a massively parallel approach to genetic sequence analysis that is having a major impact on biomedical research and clinical diagnostics *(1)*. These arrays are comprised of large sets of nucleic acid probe sequences immobilized in defined, addressable locations on the surface of a substrate, and are capable of acquiring unprecedented amounts of genetic information from biological samples in a single hybridization procedure. The advent of this technology has relied on developing methods of fabricating arrays with sufficiently high information content and density. Light-directed synthesis *(2–5)* has enabled the large-scale manufacture of arrays containing hundreds of thousands of oligonucleotide probe sequences on a glass "chip" about 1.6 cm². This method is used to produce high-density GeneChip® probe arrays, which are now finding widespread use in the detection and analysis of mutations and polymorphisms ("genotyping"), and in a wide range of gene expression studies.

In a process combining DNA synthesis chemistry with photolithographic techniques adapted from the semiconductor industry, 5'- or 3'-terminal protecting groups are selectively removed from growing oligonucleotide chains in predefined regions of a glass support by controlled exposure to light through photolithographic masks (**Fig. 1**). A planar glass or fused silica substrate is first covalently modified with a silane reagent to provide hydroxyalkyl groups, which serve as the initial synthesis sites. These sites are extended with linker groups protected with a photolabile-protecting group such that when specific regions of the surface are exposed to light, they are selectively "activated" for the addition of nucleoside phosphoramidite monomers. The monomers, which

From: *Methods in Molecular Biology, vol. 170: DNA Arrays: Methods and Protocols*
Edited by: J. B. Rampal © Humana Press Inc., Totowa, NJ

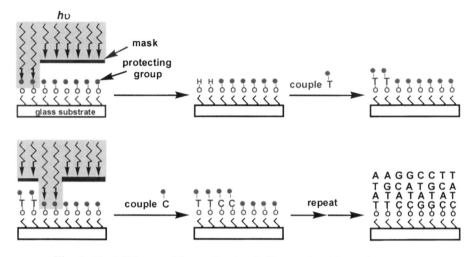

Fig. 1. Photolithographic synthesis of oligonucleotide probe arrays.

are also protected at the 5' (or 3') position with a photolabile group, are coupled to the substrate using standard phosphoramidite DNA synthesis protocols *(4)*. Cycles of photodeprotection and nucleotide addition are repeated to build the desired two-dimensional array of sequences.

The photolithographic process allows massively parallel synthesis of large sets of probe sequences, and provides a very efficient route to high-density arrays: a complete set, or any subset, of all probe sequences of length n requires, at most, $4n$ synthesis steps. For example, the complete array of all possible 10-mer probes ($>10^6$ sequences) requires only 40 cycles. Mask sets can be designed to make virtually any array of oligonucleotide probe sequences for a variety of applications. Typical arrays comprise customized sets of probes that are 20–25 bases in length. Semiautomated manufacturing techniques and lithography tools have been adapted from the microelectronics industry for large-scale commercial GeneChip® array production in a multichip wafer format.

The spatial resolution of the photolithographic process ultimately determines the maximum achievable density of the array and therefore the amount of sequence information that can be encoded on a chip of a given physical dimension. Currently, arrays made using photolithographic synthesis have individual probe features 24×24 μ on a 1.6-cm^2 chip, but further refinements in the technology will provide a resolution that will allow arrays to be fabricated with densities $>10^6$ sequences/cm^2, corresponding to feature sizes less than 10×10 μ^2.

The current methodology employs nucleoside monomers protected with a photoremovable 5'-(α-methyl-6-nitropiperonyloxycarbonyl) (MeNPOC) group *(3,4)*, which offers a number of advantages. Not insignificantly, MeNPOC

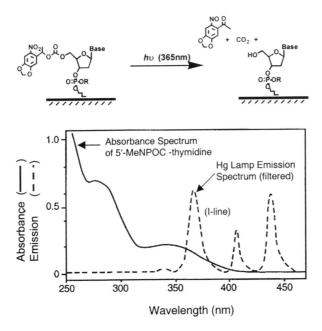

Fig. 2. Photolysis reaction of MeNPOC protecting group, and overlay of MeNPOC-nucleoside absorbance spectrum and Hg lamp emission spectrum.

phosphoramidite reagents are relatively inexpensive to prepare. Photolysis is induced by irradiation with near-ultraviolet (UV) light (ϕ_{365} ~0.05; λ_{max} ~350 nm) (**Fig. 2**) so that longer wavelengths can be used to avoid photochemical modification of the oligonucleotides, which absorb energy at below ~320 nm. The photolysis reaction involves an intramolecular redox reaction and does not require any special solvents, catalysts, or coreactants. Complete deprotection requires <1 min using filtered Hg I-line (365 ± 10 nm) emission from a commercial photolithographic exposure system. Also, photolysis "rates" are independent of the associated nucleotide base and oligomer length, which conveniently allows the use of a single exposure setting for the entire process.

The average stepwise efficiency of oligonucleotide synthesis, in this format using standard bases, is in the 90–95% range. These values reflect the yield of the photochemical deprotection step after exhaustive photolysis (*4*). Subquantitative photolysis yields lead to incomplete or "truncated" probes, with the desired full-length sequences representing, in the case of 20-mer probes, approx 10% of the total. However, this has a relatively minor impact on the performance characteristics of the arrays when they are used for hybridization-based sequence analysis. First, there is ample density of surface synthesis sites using the silanating agent described above (approx 120 pmol/cm², unpublished observation) so that the absolute amount of completed probes on the support remains high. Increasing the synthesis yield through alternate chemistries or processes available does increase the surface concentration of full-length probes. However, this can actually decrease hybridization signal intensity owing to the steric/electrostatic repulsive effects that result when oligonucleotides are too closely spaced on the support. There is an optimum probe density for maximum hybridization signal. Second, array hybridizations are typically carried out under stringent conditions whereby hybridization to significantly shorter ($<n-4$) oligomers is relatively minor. These factors, combined with the use of comparative intensity algorithms for data analysis *(5)*, allow highly accurate sequence information to be "read" from these arrays with single-base resolution.

Alternative photolabile-protecting groups have been described that also may be applicable to light-directed DNA array synthesis, but these have not seen extensive use *(6–11)*. To achieve higher photolysis rates, synthesis yields, and spatial resolution, we have also developed photolithographic methods for fabricating DNA arrays that exploit polymeric photoresist films as the photoimageable component *(12–14)*. One such approach uses a polymer film containing a chemically amplified photoacid generator, wherein exposure to light creates localized acid development adjacent to the substrate surface, resulting in direct removal of 4,4'-dimethoxytrityl (DMT)-protecting groups from the oligonucleotide chains. Such high-resolution photoresist-based processes will enable production of oligonucleotide arrays with features on the order of 5 μ in size.

In this chapter, we describe methods for the synthesis of photolabile MeNPOC nucleoside phosphoramidite building blocks for standard 3'–5' as well as 5'–3' photolithographic synthesis, and for "modified" 2'-*O*-methyl and 2,6-diaminopurine nucleosides that can be used to improve hybridization affinities of AT-probe sequences. We also outline the photolithographic synthesis method, general protocols for determining photochemical deprotection rates and yields for oligonucleotide synthesis based on surface fluorescence, as well as procedures based on hybridization for comparing the array performance characteristics of new chemistries and protocols.

2. Materials

2.1. Chemicals

1. General reagents and anhydrous solvents were obtained from Aldrich.
2. Phosgene (20% [w/v]) in toluene (Fluka).
3. Thymidine, 2'-deoxyinosine, N^2-phenoxyacetyl-2'-deoxyguanosine, N^2-isobutyryl-2'-deoxyguanosine, N^4-isobutyryl-2'-deoxycytidine, N^6-phenoxyacetyl-2'-deoxy adenosine, and diisopropylammonium tetrazolide (Chem-Impex, Wood Dale, IL).
4. 3,4-Methylenedioxyacetophenone (Aldrich).

5. 2,6-Diaminopurine-2'-deoxyriboside, 2,6-diaminopurine-2'-*O*-methylriboside, 5'-DMT-2'-*O*-methyl(N^6-benzoyl)adenosine, and 5'-DMT-2'-*O*-methyl-5-methyluridine (Reliable Biopharmaceuticals, St. Louis, MO).

6. 5'-DMT-thymidine, 5'-DMT-N^2-isobutyryl-2'-deoxyguanosine, 5'-DMT-N^4-isobutyryl-2'-deoxycytidine, N^6-benzoyl-2'-deoxyadenosine, DMT-[OCH_2CH_2]$_6$O-CED and 2-cyanoethyl-*N,N,N',N'*-tetraisopropylphosphorodiamidite (ChemGenes, Waltham, MA).

7. *N,N*-(diisopropyl)dimethylphosphoramidite ("AmCAP") (Toronto Research Chemicals, Toronto, Canada).

8. Fluorescein phosphoramidite (Fluoreprime™) and 5'-DMT-2'-deoxynucleoside-3'-phosphoramidites (Amersham-Pharmacia Biotech).

9. N,N-*bis*(2-hydroxyethyl)-3-aminopropyltriethoxysilane (Gelest, Tullytown, PA).

10. Ancillary DNA synthesis reagents (Glen Research, Sterling, VA).

11. Silica gel: 60-Å pore size, 230–400 mesh for flash chromatography (E. Merck).

12. Glass microscope slides, soda-lime, $2 \times 3 \times 0.027$ in. (Erie Scientific).

13. Nanostrip (Cyantek, Fremont, CA).

14. Hybridization buffer: 6X SSPE (0.9 *M* NaCl, 60 m*M* NaH_2PO_4, and 6 m*M* EDTA), pH 7.5.

2.2. Apparatus

1. High-performance liquid chromatography (HPLC) analyses were performed on Beckman System Gold using absorbance detection at 260 nm, and a Beckman reverse-phase column (5 µm C18-silica, 4 mm id × 250 mm length) eluted with a linear gradient of acetonitrile in 0.1 *M* aqueous triethylammonium acetate, pH 7.2, at a flow rate of 1 mL/min.

2. UV-visible spectra were obtained on a Varian Cary 3E spectrophotometer.

3. ^{1}H and ^{31}P nuclear magnetic resonance (NMR) spectra were recorded on a Varian Gemini-400 spectrometer.

4. Mass spectra were recorded on the following instruments: electrospray ionization (ESI-MS), Hewlett Packard HP59987A; electron impact ionization (EI-MS), Hewlett Packard HP5989B; and positive ion fast atom bombardment (FAB-MS), MicroMass ZABZ-EQ.

5. Elemental analyses were performed by Quantitative Technologies, Whitehouse, NJ.

6. Light source equipped with Ushio model ush508sa super-high-pressure mercury lamp and dichroic reflectors to provide output in the near-UV spectral range (model 87330, Oriel Instruments, Stratford, CT).

3. Methods

3.1. Synthesis of Photolabile MeNPOC Phosphoramidite Reagents

3.1.1. PAC-Protected 2,6-Diaminopurine Nucleosides

2,6-*bis*(phenoxyacetylamino)purine-2'-deoxyribonucleoside and 2,6-*bis*(phenoxyacetylamino)purine 2'-*O*-methylribonucleoside were prepared by the following general procedure. All intermediates were characterized by ^{1}H NMR.

The 3',5'-Tetraisopropyldisiloxylnucleosides were prepared according to the method in **ref. 15** and purified by flash chromatography. The 3',5'-TIPS-nucleoside (52 mmol) was then dried by coevaporation (2X pyridine), and dissolved in 520 mL of anhydrous pyridine. Phenoxyacetic anhydride (250 mmol, 5 Equiv.) was added and the reaction stirred overnight. The mixture was evaporated to an oil and coevaporated twice from toluene. The residue was dissolved in dichloromethane (DCM), washed twice with 5% $NaHCO_3$, and once with brine, dried with $MgSO_4$, and evaporated to a foam. The crude 2,6-*bis*(phenoxyacetylamino)purine-3',5'-tetraisopropyldisiloxyl-nucleosides were purified by flash chromatography in DCM-MeOH (yield: 80–90%).

The PAC/TIPS-protected nucleoside (40 mmol) was dried by coevaporation (2X toluene) and then stirred with 5 equiv. of [$Et_3N \cdot 3HF$] in 400 mL dry THF overnight *(16)*. The solvent was evaporated and the residue resuspended in 150 mL of anhydrous tetrahydrofuran (THF) and triturated with 2.5 L of ethyl ether for 1 h. The resulting powdery solid was collected by filtration, washed with ethyl ether, and dried. It was then suspended in 1 L of H_2O with stirring for 15 min, filtered, washed, and dried under high vacuum to obtain the 2,6-*bis*(phenoxyacetylamino)purine nucleosides in 65–85% yield.

3.1.2. Synthesis of MeNPOC-Chloride

3.1.2.1. METHYL(3,4-METHYLENEDIOXY-6-NITROPHENYL)KETONE

A solution of 410 g (2.5 mol) of 3,4-methylenedioxyacetophenone in 1600 mL of glacial acetic acid was added dropwise over 45 min to 3400 mL of cold 70% HNO_3. The solution was stirred continuously and maintained at 3–5°C throughout the addition, and for another 60 min afterward. The mixture was allowed to stir at ambient temperature for another 2 h, and then slowly was poured into 10 L of crushed ice. The resulting yellow solid was collected by filtration, washed with water, dried under vacuum, and recrystallized from THF-hexane to afford 345 g (66%) of product. ^{1}H NMR ($CDCl_3$, 400 MHz): δ 7.55 (s, 1H), 6.75 (s, 1H), 6.18 (s, 2H), 2.49 (s, 3H). ESI-MS: 210 ($M+H^+$), 232 ($M+Na^+$). Anal. Calc. for $C_9H_7NO_5$: C, 51.68; H, 3.37; N, 6.70. Found: C, 51.56; H, 3.21; N, 6.40. The product was contaminated with a small amount (3–5% by HPLC) of an inert byproduct, 1,2-methylenedioxy-4-nitrobenzene. ^{1}H NMR (dimethyl sulfoxide [DMSO]-d_6, 400 MHz): δ 7.93 (d, $J = 10$ Hz, 1H), 7.80 (s, 1H), 7.17 (d, $J = 10$ Hz, 1H), 6.30 (s, 2H).

3.1.2.2. (R,S)-1-(3,4-METHYLENEDIOXY-6-NITROPHENYL)ETHANOL

Sodium borohydride (55 g; 1.45 mol) was added over 60 min to a cold, stirring suspension of 3,4-methylenedioxy-6-nitroacetophenone (690 g; 3.3 mol) in 3200 mL of methanol, maintaining a temperature of 5–15°C with external cooling. Stirring was continued at ambient temperature until the reaction was complete (2 to 3 h), at which time the mixture was combined with 1600 mL of saturated aqueous ammonium chloride. The resulting suspension was extracted three times with CH_2Cl_2, and the combined extracts were washed with brine, dried with $MgSO_4$, and evaporated to give 680 g (98%) of product after drying under vacuum. ^{1}H NMR ($CDCl_3$, 200 MHz): δ 7.46 (s, 1H), 7.27 (s, 1H), 6.11 (s, 2H), 5.46 (quartet, J = 6.5 Hz, 1H), 1.54 (d, J = 6.5 Hz, 3H).

3.1.2.3. (*R*,*S*)-1-(3,4-METHYLENEDIOXY-6-NITROPHENYL)ETHYL CHLOROFORMATE (MeNPOC-Cl)

A solution of phosgene in toluene (1250 mL of 20% [w/v]; 2.4 mol) was added to a stirring suspension of (*R*,*S*)-1-(3,4-methylenedioxy-6-nitrophenyl)ethanol (211 g; 1.0 mol) in 500 mL of dry THF. Stirring was continued at ambient temperature under argon until the reaction was complete (36–48 h). Excess phosgene was removed under low vacuum with an aqueous NaOH trap, before the remaining solvents were removed on a rotary evaporator. The residue was triturated with hexane to obtain a solid brown cake, which was then recrystallized from THF-hexane to afford 205 g of product as a light-brown powder which was at least 95% pure according to ^{1}H NMR: ($CDCl_3$, 200 MHz); δ 7.52 (s, 1H), 7.05 (s, 1H), 6.47 (quartet, J = 6.5 Hz, 1H), 6.16 (s, 2H), 1.78 (d, J = 6.5 Hz, 3H). The material was stored desiccated at –20°C.

3.1.3. 5'-MeNPOC-2'-Deoxyribonucleosides

Base-protected nucleosides (90 mmol) were dried by coevaporating three times with 250 mL of anhydrous pyridine, dissolved or suspended in 300 mL of anhydrous pyridine under argon and then cooled to –40°C in a dry ice–acetonitrile bath. A solution of 27.5 g (100 mmol) of MeNPOC-Cl in 100 mL dry CH_2Cl_2 was then added dropwise with stirring. After 30 min, the cold bath was removed, and the solution was allowed to stir overnight at room temperature. After evaporating the solvents, the crude material was taken up in EtOAc and extracted with water and brine. The organic phase was dried over Na_2SO_4 and evaporated to obtain yellow foam. The crude products were generally purified by flash chromatography on silica gel (0–6% MeOH gradient in CH_2Cl_2 or 1:1 CH_2Cl_2-EtOAc), except MeNPOC-dI, which was recrystallized from DCM. Yields of purified 5'-*O*-MeNPOC-nucleoside (mixture of diastereomers) were in the range of 65–85%, with purity >96% as determined by reverse-phase HPLC (RP-HPLC) (15–100% CH_3CN/15 min gradient), and ^{1}H NMR.

3.1.4. Synthesis of 3'-MeNPOC-2'-Deoxynucleosides

5'-DMT-base-protected nucleosides (50 mmol) were dried by coevaporating three times with 150 mL of anhydrous pyridine. The nucleoside was then dissolved or suspended in 150 mL of an anhydrous mixture of pyridine and DCM (2:1 [v]) under argon, and cooled to –40°C (dry ice-CH_3CN). A solution of 15 g (55 mmol) of MeNPOC-Cl in 40 mL of dry DCM was then added dropwise with stirring. After 30 min, the cold bath was removed, and the solution was allowed to stir overnight at room temperature (thin-layer chromatography [TLC]: DCM/EtOAc). After removing the solvents, the crude material was taken up in EtOAc and washed with water and brine. The organic phase was dried over Na_2SO_4 and evaporated to obtain a yellow foam.

The crude 5'-DMT-3'-MeNPOC-nucleoside (from 50 mmol rxn) was detritylated by stirring in 1 L of 3% trichloroacetic acid in DCM-MeOH (85:15) at ambient temperature. The reaction was monitored by TLC (~1:1 DCM-EtOAc), and when detritylation was complete (approx 60–120 min), the mixture was transferred to a separatory funnel and washed twice with 250 mL of saturated aqueous $NaHCO_3$ (CO_2 was evolved) and then once with saturated NaCl. The organic phase was dried over Na_2SO_4 and evaporated to dryness. The residue was dried by twice adding and evaporating dry acetonitrile and then purified by flash chromatography on a 9.0 wide × 12 high column of silica gel eluted with EtOAc + acetone (0 to ~50% gradient elution). Pure product (≥95% by HPLC: 15–100% CH_3CN/15 min gradient, and ^1H NMR) was obtained in ~65% overall yield.

3.1.5. 5'-MeNPOC-2'-O-Methylribonucleosides

The 5'-MeNPOC-2'-O-methyl-(N^6-bz)adenine-, (N^2,N^6-di-pac)-2.6-diaminopurine-, and 5-methyluracil-ribonucleosides were prepared according to the following general sequence (*see* **Fig. 3**).

3.1.5.1. 3'-TBDMS-5'-DMT-(N-PROTECTED)-2'-O-METHYLRIBONUCLEOSIDES

The 5'-DMT-(N-protected)-2'-O-methylribonucleosides (9 mmol) were dried by coevaporation (2X pyridine), dissolved in pyridine (90 mL), and cooled to –20°C. The *t*-butyldimethysilyltrifluoromethanesulfonate (9.9 mmol, 1.1 Eq.) was added dropwise and the reaction allowed to stir at –20°C for 3 h before warming to room temperature. On completion (by TLC), the mixture was evaporated to half-volume and poured into 500 mL of DCM. The organic layer was washed with ice-cold $NaHCO_3$ (saturated) and then brine and dried with Na_2SO_4. After evaporating the solvents, the crude material was dried by coevaporation (3X toluene), placed under high vacuum (yield = ~92%), and used without further purification.

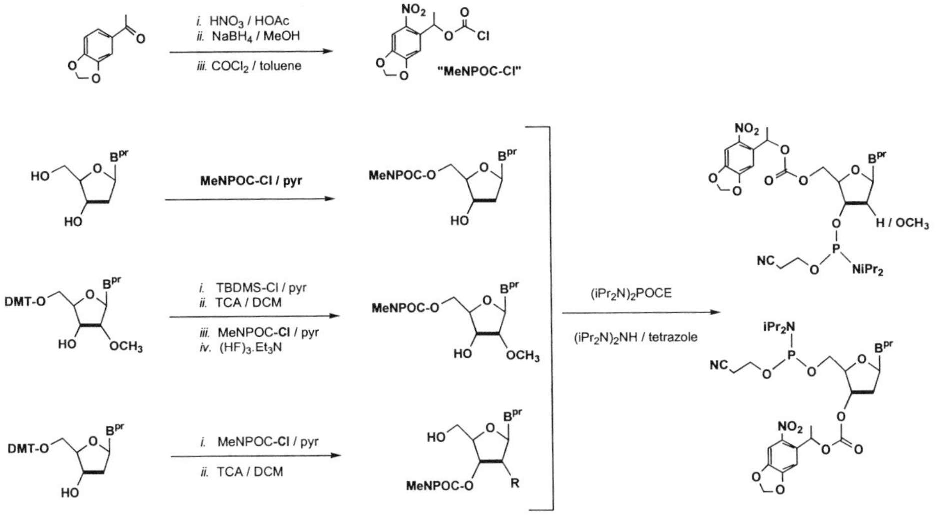

Fig. 3. Synthesis of 3'-, and 5'-MeNPOC nucleoside phosphoramidites.

3.1.5.2. 3'-TERT-BUTYLDIMETHYSILYL-(*N*-PROTECTED)-2'-*O*-METHYLRIBOSIDE

To a stirring solution of the crude 3'-TBDMS-5'-DMT-(*N*-protected)-2'-*O*-methylribonucleoside (5.2 mmol) in DCM (100 mL) was added a solution of 3% TCA acid in DCM (60 mL) and MeOH (15 mL). After 1 h, detritylation was complete and the reaction was quenched by washing twice with ice–saturated NaHCO$_3$, followed by brine. The organic phase was evaporated to an oil and dried by coevaporation with CH$_3$CN.

The 3'-silylated purine nucleosides were purified by flash chromatography: (*N*^{2},*N*^{6}-di-pac)-2,6-diaminopurine-2'-*O*-methylribonucleoside, DCM-MeOH (1–5%); (*N*^{6}-Bz)-2'-*O*-methylriboadenosine, DCM-EtOAc 7:3 (isocratic). The 3'-silylated 2'-*O*-methyl-5-methyluridine was purified by dissolving in a minimal amount of DCM and precipitating with 600 mL of Et$_2$O/hexanes (1:5) (yield: 45–95%).

3.1.5.3. 5'-MeNPOC-(*N*-PROTECTED)-2'-*O*-METHYLRIBONUCLEOSIDES

The 3'-TBDMS-(*N*-protected)-2'-*O*-methylribonucleosides were treated with MeNPOC-Cl (1.1 Eq.), as described in **Subheading 3.1.3.**, to give the 5'-MeNPOC-(*N*-protected)-2'-*O*-methylribonucleosides in 84–95% yield after purification. The silyl ethers were then cleaved by stirring with 5 eqiv. of Et$_3$N·(HF)$_3$ in dry THF overnight (*16*). After evaporating the solvent, the 5'-MeNPOC-2'-*O*-methyl-5-methyluridine and the 5'-MeNPOC-*N*^{6}-benzoyl-2'-

O-methyladenosine products were purified by trituration with ether/hexane as described in **Subheading 3.1.1.** The 5'-MeNPOC-*N,N*-(PAC)$_2$-2,6-diaminopurine-2'-*O*-methylribonucleoside was purified by flash chromatography (0–5% MeOH in 1:1 EtOAc-DCM) to >96% purity, as determined by RP-HPLC (15–100% CH$_3$CN/15 min gradient) and ^{1}H NMR. Yields of 85–92% were obtained.

3.1.6. Synthesis of MeNPOC-Nucleoside-(2-cyanoethyl)-N,N-*diisopropylphosphoramidites*

All MeNPOC-deoxynucleosides were phosphitylated using the procedure of Barone et al. *(17)*. The nucleoside (50 mmol) was combined with 16.6 g (55 mmol) of 2-cyanoethyl-*N,N,N',N'*-tetraisopropylphosphorodiamidite and 4.3 g (25 mmol) of diisopropylammonium tetrazolide in 250 mL of dry DCM under argon at ambient temperature. Stirring was continued until TLC analysis (45:45:10 hexane:DCM:Et$_3$N) indicated complete conversion (4–16 h). The reaction mixture was extracted with 200 mL of each of saturated aqueous NaHCO$_3$ and saturated brine, then dried over Na$_2$SO$_4$ and evaporated to dryness. The crude amidites (mixture of diastereomers) were purified by flash chromatography (DCM-EtOAc containing 1% triethylamine). Column fractions were assayed for purity by HPLC (40–100% CH$_3$CN over 15 min). The purified amidites were recovered by pooling and evaporating the appropriate fractions (purity by HPLC, ^{31}P-NMR ≥96%), coevaporated once with anhydrous acetonitrile, and then dried under vacuum for ~24 h. Impurities consisted of varying amounts (total 1–4%) of the hydrolyzed product (^{31}P-NMR, δ + 8.2 ppm) and the hydrolyzed phosphitylating reagent (^{31}P-NMR, δ +14.9 ppm). Yields of purified phosphoramidites were in the range of 60–90%.

3.1.7. 18-O-MeNPOC-3,6,9,12,15,18-Hexaoxaoctadec-1-yl-(2-cyanoethyl)-N,N-*diisopropylphosphoramidite (MeNPOC-HEG-CEP)*

Hexaethyleneglycol (200 g, 710 mmol) was coevaporated twice with 400 mL of dry pyridine, then dissolved in 500 mL of dry pyridine and DCM (4:6). MeNPOC-Cl (64.5 g, 236 mmol) in 500 mL of CH$_2$Cl$_2$ was then added dropwise with stirring over 2 h. After stirring for another hour, the mixture was transferred to a separatory funnel and washed three times with 150 mL of water and once with 150 mL of saturated brine. The organic phase was dried (NaSO$_4$) and evaporated to give ~125 g of crude material as an oil, which, according to HPLC analysis, consisted of an 8:2 mixture of mono- and *bis*-acylated products. The crude material was purified by flash chromatography (0–5% MeOH in DCM-EtOAc) to give 80 g of 18-*O*-MeNPOC-3,6,9,12,15,18-hexaoxaoctadecan-1-ol (≥98% purity by RP-HPLC). ^{1}H NMR (CDCl$_3$, 400 MHz): δ 7.50 (s, 1H, H$_{Ar-MeNPOC}$), 7.09 (s, 1H, H$_{Ar-MeNPOC}$), 6.27 (quartet, *J* = 6.5 Hz, 1H, H$_{ArCH-MeNPOC}$), 6.12 (br s, 2H, H$_{OCH2O-MeNPOC}$), 4.80 (br s, 1H, H$_{OH1}$), 4.31–

4.19 (m, 2H, H_{C17}), 3.73–3.59 (m, 22H, H_{C1-16}), 1.64 (d, J = 8Hz, 3H, $H_{CH3-MeNPOC}$). ESI-MS: 520 (M+H$^+$), 537 (M+H$_2$O$^+$), 542 (M+Na$^+$).

The MeNPOC-HEG-OH (80 g, 154 mmol) was co-evaporated three times with 300 mL of dry toluene, and combined with 2-cyanoethyl-N,N,N',N'-tetraisopropylphosphorodiamidite (55.7 g, 185 mmol) and diisopropyl-ammoniumtetrazolide (6.4 g, 77 mmol) in 600 mL of dry DCM under argon at room temperature. The solution was stirred overnight, then transferred to a separatory funnel and washed twice with saturated aqueous NaHCO$_3$ and once with saturated brine (400 mL each). After drying with Na$_2$SO$_4$, the organic phase was evaporated and the crude product was purified by flash chromatography (0–50% EtOAc in DCM/0.5% triethylamine) to give 87 g (78.5%) of MeNPOC-HEG-CEP (purity: 98% pure by HPLC, 100% by ^{31}P NMR). ^{1}H NMR (CDCl$_3$, 400 MHz): δ 7.50 (s, 1H, $H_{Ar-MeNPOC}$), 7.09 (s, 1H, $H_{Ar-MeNPOC}$), 6.27 (quartet, J = 6.5 Hz, 1H, $H_{ArCH-MeNPOC}$), 6.12 (br s, 2H, $H_{OCH2O-MeNPOC}$), 4.30–4.18 (m, 2H, H_{C17}), 3.90–3.80 (m, 3H, $H_{C1, CE-a}$), 3.76–3.58 (m, 23H, $H_{C2-16, CH-iPr, CE-a}$), 2.65 (t, J = 8Hz, H_{CE-b}), 1.65 (d, J = 8Hz, 3H, $H_{CH3-MeNPOC}$), 1.21–1.16 (m, 12H, $H_{CH3-iPr}$). ^{31}P NMR (CDCl$_3$, 162 MHz): δ +148.7. ESI-MS: 720 (M+H$^+$), 742 (M+Na$^+$), 821 (M+Et$_3$NH$^+$). Anal. calc. for C$_{31}$H$_{51}$N$_3$O$_{14}$P: C, 51.66; H, 7.13; N, 5.83; P, 4.30. Found: C, 51.70; H, 6.82; N, 5.73; P, 4.04.

3.1.8. Analytical Data for Photolabile MeNPOC Nucleoside Phosphoramidite Reagents

3.1.8.1. 5'-O-MeNPOC-N^2-PHENOXYACETYL-2'-DEOXYADENOSINE-3'-O-(2-CYANOETHYL)-N,N-DIISOPROPYLPHOSPHORAMIDITE (5'-MeNPOC-[PAC]dA-CEP)

^{1}H NMR (CDCl$_3$, 400 MHz): δ 9.38 (br s, 1H, H_{N6}), 8.79 (s, 1H, H_{C2}), 8.28, 8.22 (2s, 1H, H_{C8}), 7.47, 7.46 (2s, 1H, $H_{Ar-MeNPOC}$), 7.37–7.32 (m, 2H, H_{Ar-PAC}), 7.10-7.02 (m, 4H, $H_{Ar-PAC, MeNPOC}$), 6.53–6.47 (m, 1H, $H_{C1'}$), 6.31–6.24 (m, 1H, $H_{CH-MeNPOC}$), 6.10-6.04 (m, 2H, $H_{OCH2O-MeNPOC}$), 4.93, 4.89 (2br s, 2H, H_{PAC}), 4.85–4.70 (br d, 1H, $H_{C3'}$), 4.51–4.30 (m, 2H, $H_{C5',4'}$), 3.93–3.85 (br m, 1H, H_{CE-a}), 3.81–3.73 (br m, 1H, H_{CE-a}), 3.68–3.58 (br m, 2H, H_{CH-iPr}), 3.02–2.88 (m, 1H, $H_{C2'a}$), 2.80–2.62 (br m, 3H, $H_{C2'b, CE-b}$), 1.66–1.62 (m, 3H, $H_{CH3-MeNPOC}$), 1.23–1.14 (m, 12H, $H_{CH3-iPr}$). ^{31}P NMR (CDCl$_3$, 162 MHz): δ +149.6. FAB-MS: 823.2 (M+H$^+$). Anal. Calc. for C$_{37}$H$_{43}$N$_8$O$_{12}$P: C, 54.01; H, 5.27; N, 13.62; P, 3.76. Found: C, 54.24; H, 5.40; N, 13.48; P, 4.12.

3.1.8.2. 5'-O-MeNPOC-N^2-ISOBUTYRYL-2'-DEOXYGUANOSINE-3'-O-(2-CYANOETHYL)-N,N-DIISOPROPYLPHOSPHORAMIDITE (5'-MeNPOC-[IBU]dG-CEP)

^{1}H NMR (CDCl$_3$, 400 MHz): δ 9.01, 8.92, 8.84, 8.73 (4 br s, 1H, H_{N2}), 7.88, 7.78 (2d, J_{app} = 12 Hz, 1H, H_{C8}), 7.48–7.46 (2s, 1H, $H_{Ar-MeNPOC}$), 7.03, 7.00,

6.92, 6.88 (4s, 1H, $H_{Ar-MeNPOC}$), 6.32–6.28 (m, 1H, $H_{C1'}$), 6.22–6.03 (m, 3H, $H_{ArCH, OCH2O-MeNPOC}$), 4.85–4.30 (br m, 4H, $H_{C3',5',4'}$), 3.96–3.88 (br m, 1H, H_{CE-a}), 3.80–3.72 (br m, 1H, H_{CE-a}), 3.68–3.58 (br m, 2H, H_{CH-iPr}), 2.98–2.44 (mm, 5H, $H_{C2'a, C2'b, CE-b,CH-iBu}$), 1.70 (br s, 1H, H_{N5}), 1.65–1.61 (m, 3H, $H_{CH3-MeNPOC}$), 1.28–1.16 (m, 18H, $H_{CH3-iPr, iBu}$). ^{31}P NMR (CDCl$_3$, 162 MHz): δ +149.2. ESI-MS: 775 (M+H$^+$), 797 (M+Na$^+$), 876 (M+Et$_3$NH$^+$). Anal. calc. for $C_{33}H_{43}N_8O_{12}P$: C, 51.16; H, 5.60; N, 14.46. Found: C, 51.08; H, 5.57; N, 13.92.

3.1.8.3. 5'-*O*-MᴇNPOC-*N²*-ᴘʜᴇɴᴏxʏᴀᴄᴇᴛʏʟ-2'-ᴅᴇᴏxʏɢᴜᴀɴᴏsɪɴᴇ-3'-*O*-(2-ᴄʏᴀɴᴏ-ᴇᴛʜʏʟ)-*N,N*-ᴅɪɪsᴏᴘʀᴏᴘʏʟᴘʜᴏsᴘʜᴏʀᴀᴍɪᴅɪᴛᴇ (5'-MᴇNPOC-[ᴘᴀᴄ]ᴅG-CEP)

^{1}H NMR (CDCl$_3$, 400 MHz): δ 7.88, 7.78 (2d, J_{app} = 10 Hz, 1H, H_{C8}), 7.42–7.32 (m, 4H, H_{Ar}), 7.11-6.86 (m, 3H), 6.32–5.97 (m, 4H, $H_{ArCH, OCH2O-MeNPOC}$), 4.75–4.26 (br m, 7H, $H_{C3',5',4', PAC}$), 3.93–3.85 (br m, 1H, H_{CE-a}), 3.80–3.72 (br m, 1H, H_{CE-a}), 3.68–3.58 (br m, 2H, H_{CH-iPr}), 2.90–2.70 (br m, 1H, $H_{C2'a}$), 2.70–2.62 (br m, 2H, H_{CE-b}), 2.62–2.35 (br m, 1H, $H_{C2'b}$), 1.75 (br s, 2H, H_{NH}), 1.62, 1.59 (2d, 3H, $H_{CH3-MeNPOC}$), 1.21 (d, J = 8 Hz, 12H, $H_{CH3-iPr}$). ^{31}P NMR (CDCl$_3$, 162 MHz): δ +149.5. FAB-MS: 839.2 (M+H$^+$). Anal. calc. for $C_{37}H_{43}N_8O_{13}P$: C, 52.98; H, 5.17; N, 13.36; P, 3.69. Found: C, 53.12; H, 5.30; N, 13.51; P, 3.74.

3.1.8.4. 5'-*O*-MᴇNPOC-2'-ᴅᴇᴏxʏɪɴᴏsɪɴᴇ-3'-*O*-(2-ᴄʏᴀɴᴏᴇᴛʜʏʟ)-*N,N*-ᴅɪɪsᴏᴘʀᴏᴘʏʟᴘʜᴏsᴘʜᴏʀᴀᴍɪᴅɪᴛᴇ (5'-MᴇNPOC-ᴅI-CEP)

^{1}H NMR (CDCl$_3$, 400 MHz): δ 8.23, 8.20 (2s, 1H, H_{C2}), 8.17, 8.14 (2s, 1H, H_{C8}), 7.59, 7.58 (2s, 1H, $H_{Ar-MeNPOC}$), 7.14, 7.13 (2s, 1H, $H_{Ar-MeNPOC}$), 6.54–6.49 (m, 2H, $H_{1'}$), 6.41-6.35 (m, 1H, $H_{ArCH-MeNPOC}$), 6.27-6.26 (4s, 2H, $H_{OCH2O-MeNPOC}$), 4.91–4.79 (m, 1H, $H_{C3'}$), 4.61–4.40 (m, 2H, $H_{C5',4'}$), 4.04–3.96 (br m, 1H, H_{CE-a}), 3.93–3.83 (br m, 1H, H_{CE-b}), 3.81–3.70 (br m, 2H, H_{CH-iPr}), 3.20 (br s, 1H, H_{N1}), 3.02–2.89 (m, 1H, $H_{C2'a}$), 2.87–2.70 (m, 3H, $H_{C2'b, CE-b}$), 1.76 (d, J = 8 Hz, 3H, $H_{CH3-MeNPOC}$), 1.35–1.30 (m, 12H, $H_{CH3-iPr}$). ^{31}P NMR (CDCl$_3$, 162 MHz): δ +149.7. ESI-MS: 690 (M+H$^+$), 712 (M+Na$^+$), 791 (M+Et3NH$^+$). Anal. calc. for $C_{29}H_{36}N_7O_{11}P$: C, 50.50; H, 5.26; N, 14.22; P, 4.49. Found: C, 50.28; H, 5.21; N, 14.06; P, 4.13.

3.1.8.5. 5'-*O*-MᴇNPOC-*N²*-ɪsᴏʙᴜᴛʏʀʏʟ-2'-ᴅᴇᴏxʏᴄʏᴛᴏsɪɴᴇ-3'-*O*-(2-ᴄʏᴀɴᴏᴇᴛʜʏʟ)-*N,N*-ᴅɪɪsᴏᴘʀᴏᴘʏʟᴘʜᴏsᴘʜᴏʀᴀᴍɪᴅɪᴛᴇ (5'-MᴇNPOC-[ɪBᴜ]ᴅC-CEP)

^{1}H NMR (CDCl$_3$, 400 MHz): δ 8.96–8.83 (br m, 1H, H_{N4}), 7.99–7.93 (m, 1H, H_{C6}), 7.49–7.47 (m, 1H, $H_{Ar-MeNPOC}$), 7.43–7.39 (m, 1H, H_{C5}), 7.02, 7.00 (m, 1H, $H_{Ar-MeNPOC}$), 6.32–6.21 (m, 2H, $H_{C1', ArCH-MeNPOC}$), 6.18–6.10 (m, 2H, $H_{-OCH2O-MeNPOC}$), 4.53–4.25 (m, 4H, $H_{C5', 3', 4'}$), 3.90–3.80 (br m, 1H, H_{CE-a}),

3.80–3.70 (br m, 1H, H_{CE-a}), 3.65–3.54 (br m, 2H, H_{CH-iPr}), 2.80–2.62 (m, 3H, $H_{CH-iBu, CE-b}$), 2.33–2.26 (m, 1H, $H_{C2'a}$), 2.20–2.10 (m, 1H, $H_{C2'b}$), 1.60 (m, 3H, $H_{CH3-MeNPOC}$), 1.24–1.13 (d, 18H, $H_{CH3-iPr, iBu}$). ^{31}P NMR (CDCl$_3$, 162 MHz): δ +149.9. ESI-MS: 735 (M+H$^+$), 757 (M+Na$^+$), 836 (M+Et$_3$NH$^+$). Anal. calc. for C$_{32}$H$_{43}$N$_6$O$_{12}$P: C, 52.31; H, 5.90; N, 11.44; P, 4.22. Found: C, 52.10; H, 5.63; N, 11.03; P, 4.05.

3.1.8.6. 5'-O-MeNPOC-THYMIDINE-3'-O-(2-CYANOETHYL)-N,N-DIISOPROPYLPHOSPHORAMIDITE (5'-MeNPOC-T-CEP)

^{1}H NMR (CDCl$_3$, 400 MHz): δ 7.49 (br s, 1H, $H_{Ar-MeNPOC}$), 7.34, 7.32, 7.30, 7.28 (4s, 1H, H_{C6}), 7.01 (br s, 1H, $H_{Ar-MeNPOC}$), 6.35–6.28 (m, 2H, $H_{C1', CH-MeNPOC}$), 6.14–6.09 (m, 2H, $H_{OCH2O-MeNPOC}$), 4.57-4.30 (m, 1H, $H_{C5', C3'}$), 4.23, 4.18 (2br m, 1H, $H_{C4'}$), 3.91–3.80 (br m, 1H, H_{CE-a}), 3.78–3.68 (br m, 1H, H_{CE-a}), 3.66–3.56 (br m, 2H, H_{CH-iPr}), 2.68–2.60 (br m, 2H, H_{CE-b}), 2.57–2.40 (m, 1H, $H_{C2'a}$), 2.28–2.10 (m, 1H, $H_{C2'b}$), 1.90 (s, 3H, H_{CH3-C5}), 1.67 (br d, $J \sim 8.0$ Hz, 3H, $H_{CH3-MeNPOC}$), 1.21(d, 12H, $H_{CH3-iPr}$). ^{31}P NMR (CDCl$_3$, 162 MHz): δ +149.5. FAB-MS: 680.2 (M+H$^+$). Anal. calc. for C$_{29}$H$_{38}$N$_5$O$_{12}$P: C, 51.25; H, 5.64; N, 10.30; P, 4.56. Found: C, 51.19; H, 5.80; N, 10.48; P, 4.68.

3.1.8.7. 5'-MeNPOC-2,6-BIS(PHENOXYACETYLAMINO)PURINE-2'-DEOXYRIBOSIDE-3'-O-(2-CYANOETHYL)-N,N-DIISOPROPYLPHOSPHORAMIDITE (5'-MeNPOC-[PAC]dD-CEP)

^{1}H NMR (DMSO, 400 MHz): δ 8.55-8.51 (m, 1H, H_{C8}), 7.59, 7.56 (2s, 1H, $H_{Ar-MeNPOC}$), 7.31–7.23 (m, 4H, H_{Ar-PAC}), 7.13, 7.10 (2s, 1H, $H_{Ar-MeNPOC}$) 7.0–6.90 (m, 6H, H_{Ar-PAC}), 6.45–6.37 (m, 1H, $H_{C1'}$), 6.25–6.20 (m, 2H, $H_{OCH2O-MeNPOC}$) 6.04–5.98 (br q, 1H, $H_{CH-MeNPOC}$), 5.19, (br s, 2H, H_{PAC}), 5.08 (br s, 2H, H_{PAC}), 4.80–4.70 (br m, 1H, $H_{C3'}$), 4.44–4.13 (m, 3H, $H_{C5',4'}$), 3.84–3.68 (br m, 2H, H_{CE-a}), 3.65–3.56 (br m, 2H, H_{CH-iPr}), 3.12–3.00 (m, 1H, $H_{C2'a}$), 2.81–2.76 (br m, 2H, H_{CE-b}), 2.59–2.50 (br m, 1H) 1.57–1.52 (m, 3H), 1.20–1.12 (m, 12H). ^{31}P NMR (CDCl$_3$, 162 MHz): δ +150.0,149.8.

3.1.8.8. 5'-MeNPOC-2,6-BIS(PHENOXYACETYLAMINO)PURINE 2'-O-METHYLRIBOSIDE-3'-O-(2-CYANOETHYL)-N,N-DIISOPROPYLPHOSPHORAMIDITE (5'-MeNPOC-2'-OMe-[PAC]D-CEP)

^{1}H NMR (DMSO, 400 MHz): δ 8.60–8.53 (3s, 1H), 7.61–7.55 (m, 1H), 7.32–7.23 (m, 4H), 7.17-, 7.13 (m, 1H) 7.0–6.91 (m, 6H), 6.30–6.18 (m, 2H), 6.10–5.98 (m, 2H, $H_{C1'}$), 5.22–5.07, (br 3s, 4H), 4.85–4.60 (br m, 2H, $H_{C2' HC3'}$), 4.45–4.18 (m, 3H, $H_{C5',4'}$), 3.86–3.76 (br m, 2H, H_{CE-a}), 3.67–3.55 (br m, 2H, H_{CH-iPr}), 3.4–3.32 (m, 3H, $H_{C2'-OCH3}$) 2.82–2.75 (m, 2H, H_{CE-b}), 1.61–1.56 (m, 3H, $H_{CH3-MeNPOC}$), 1.22–1.10 (m, 12H, $H_{CH3-iPr}$). ^{31}P NMR (CDCl$_3$, 162 MHz): δ +151.7,151.4.

3.1.8.9. 5'-MeNPOC-2'-O-methyl-5-methyluridine-3'-O-(2-cyanoethyl)-N,N-diisopropylphosphoramidite (5'-MeNPOC-2'-OMe-5MeU-CEP)

^{1}H NMR (DMSO, 400 MHz): δ 7.64–7.39 (m, 2H, H$_{Ar-MeNPOC, HC6}$), 7.29–7.11 (2d, 1H, H$_{Ar-MeNPOC}$), 6.27–6.19 (m, 2H, H$_{OC\underline{H}2O-MeNPOC}$) 6.08–6.01 (m 1H, H$_{CH-MeNPOC}$) 5.87–5.83 (m, 1H, H$_{C1'}$), 4.42–3.90 (m, 5H, H$_{C2',C3', C5',C4'}$) 3.84–3.75 (br m, 2H, H$_{CE-a}$), 3.70–3.55 (br m, 2H, H$_{CH-iPr}$), 3.4–3.32 (m, H$_{C2'-OCH3}$), 2.82–2.74 (br m, 2H, H$_{CE-b}$), 1.80–1.71(m, 3H, H$_{C5CH3}$), 1.64–1.55 (m, 3H, H$_{CH3-MeNPOC}$), 1.25–1.05 (m, 12H, H$_{CH3-iPr}$). ^{31}P NMR (CDCl$_3$, 162 MHz): δ +151.25, 151.08.

3.1.8.10. 5'-MeNPOC-N⁶-Benzoyl-2'-O-methyl-adenosine-3'-O-(2-cyanoethyl)-N,N-diisopropylphosphoramidite (5'-MeNPOC-2'-OMe-[Bz]A-CEP)

^{1}H NMR (DMSO, 400 MHz): δ 8.77-8-68 (m, 2H, H$_{C2,C8}$), 8.09–8.05 (d, 2H, H$_{Ar-Bz}$), 7.68–7.65 (m, 1H, H$_{Ar-MeNPOC}$), 7.62–7.55 (m, 2H, H$_{Ar-Bz}$), 7.29–7.15 (m, 2H, H$_{Ar-MeNPOC, Ar-Bz}$), 6.25–6.16 (m, 3H, $_{HOCH2O-, CH-MeNPOC}$) 6.06–5.99 (m, 1H, H$_{C1'}$), 4.85–4.71 (m, 2H, H$_{C2'}$ H$_{C3'}$), 4.50–4.23 (m, 3H, H$_{C5',4'}$), 3.91–3.80 (br m, 2H, H$_{CE-a}$), 3.72–3.59 (br m, 2H, H$_{CH-iPr}$), 3.45–3.36 (m, 3H, H$_{C2'-OCH3}$), 2.88–2.81 (br m, 2H, H$_{CE-b}$), 1.59–1.53 (m, 3H, H$_{CH3-MeNPOC}$), 1.27–1.12 (m, 12H, H$_{CH3-iPr}$). ^{31}P NMR (CDCl$_3$, 162 MHz): δ +151.76, 151.66, and 151.09.

3.1.8.11. 3'-O-MeNPOC-N⁶-benzoyl-2'-deoxyadenosine-5'-O-(2-cyano-ethyl)-N,N-diisopropylphosphoramidite (3'-MeNPOC-[Bz]dA-CEP)

^{1}H NMR (CDCl$_3$, 400 MHz): δ 9.50–8.95 (br s, 1H, H$_{C2}$), 8.84–8.78 (br s, 1H, H$_{C8}$), 8.47–8.43 (2s, 1H, H$_{Ar-MeNPOC}$), 8.08–7.99 (d, 1H, H$_{Ar-Bz}$), 7.65–7.58 (m, 1H, H$_{Ar-Bz}$), 7.58–7.45 (m, 2H, H$_{Ar-Bz}$), 7.09, (s, 1H, H$_{Ar-MeNPOC}$), 6.66–6.56 (m, 1H, H$_{C1'}$), 6.35–6.27 (m, 1H, H$_{CH-MeNPOC}$), 6.19–6.09 (m, 2H, H$_{OCH2O-MeNPOC}$), 5.42–5.31 (br m, 1H, H$_{C3'}$), 4.40, 4.31 (2br s, 1H, H$_{C4'}$), 4.02–3.72 (br m, 4H, H$_{C5', HCE-a}$), 3.65–3.50 (br m, 2H, H$_{CH-iPr}$), 2.94–2.82 (m, 1H, H$_{C2'a}$), 2.78–2.60 (br m, 3H, H$_{C2'b, CE-b}$), 2.72–2.63 (2s, 3H, H$_{CH3-MeNPOC}$), 1.24–1.02 (mm, 12H, H$_{CH3-iPr}$). ^{31}P NMR (CDCl$_3$, 162 MHz): δ +149.6.

3.1.8.12. 3'-O-MeNPOC-N²-isobutyryl-2'-deoxyguanosine-5'-O-(2-cyanoethyl)-N,N-diisopropylphosphoramidite (3'-MeNPOC-[iBu]dG-CEP)

^{1}H NMR (CDCl$_3$, 400 MHz): δ 8.35–8.30 (br s, 1H, H$_{N2}$), 8.05, 8.04 (2s, 1H, H$_{C8}$), 7.51 (2s, 1H, H$_{Ar-MeNPOC}$), 7.07 (s, 1H, H$_{Ar-MeNPOC}$), 6.33–6.28 (m, 1H, H$_{C1'}$), 6.26–6.18 (m, 1H, H$_{ArCH-MeNPOC}$), 6.14 (m, 2H, H$_{OCH2O-MeNPOC}$), 5.40–5.35 (br m, 1H, H$_{C3'}$), 4.30, 4.25 (2br s, 1H, H$_{C4'}$), 3.92–3.84 (br m, 2H, H$_{CE-a}$), 3.84–3.72 (br m, 2H, H$_{CH-iPr}$), 3.62–3.46 (br m, 2H, H$_{C5'}$), 3.00–2.84 (br m, 1H, H$_{CH-iBu}$), 2.80–2.75 (m, 1H, H$_{C2'a}$), 2.65 (t, 1H (J = 3Hz), H$_{CE-b}$),

2.65-2.50 (m, 1H, $H_{C2'b}$), 1.70, 1.68 (2s, 3H, $H_{CH3-MeNPOC}$), 1.28–1.08 (mm, 18H, $H_{CH3-iPr, iBu}$). ^{31}P NMR (CDCl$_3$, 162 MHz): δ +150.1–149.8.

3.1.8.13. 3'-*O*-MᴇNPOC-*N²*-ɪꜱᴏʙᴜᴛʏʀʏʟ-2'-ᴅᴇᴏxʏᴄʏᴛᴏꜱɪɴᴇ-5'-*O*-(2-ᴄʏᴀɴᴏᴇᴛʜʏʟ)-*N,N*-ᴅɪɪꜱᴏᴘʀᴏᴘʏʟᴘʜᴏꜱᴘʜᴏʀᴀᴍɪᴅɪᴛᴇ (3'-MᴇNPOC-[ɪBᴜ]ᴅC-CEP)

^{1}H NMR (CDCl$_3$, 400 MHz): δ 8.24–8.18 (br m, 1H, H_{N4}), 7.97, 7.94 (m, 1H, H_{C6}), 7.50 (br s, 1H, $H_{Ar-MeNPOC}$), 7.42–7.38 (m, 1H, H_{C5}), 7.08 (br s, 1H, $H_{Ar-MeNPOC}$), 6.32–6.26 (m, 2H, $H_{C1', ArCH-MeNPOC}$), 6.16-6.14 (m, 2H, $H_{OCH2O-MeNPOC}$), 5.25–5.22; 5.16–5.14 (2 br m, 1H, $H_{C3'}$), 4.39–4.28 (m, 1H, $H_{C4'}$), 3.94–3.70 (br m, 4H, $H_{CE-a, CH-iPr}$), 3.60–3.48 (br m, 2H, $H_{C5'}$), 2.80–2.75 (m, 1H, $H_{C2'a}$), 2.65 (t, 1H, J = 3Hz, H_{CE-b}), 2.65–2.50 (m, 1H, $H_{C2'b}$), 2.90–2.72 (br m, 1H, H_{CH-iBu}), 2.66–2.60 (m, 2H, H_{CE-b}), 2.60–2.52 (m, 1H, $H_{C2'a}$), 2.20–2.10 (m, 1H, $H_{C2'b}$), 1.69–1.66 (m, 3H, $H_{CH3-MeNPOC}$), 1.24–1.12 (m, 18H, $H_{CH3-iPr, iBu}$). ^{31}P NMR (CDCl$_3$, 162 MHz): δ +150.1, 150.0.

3.1.8.14. 3'-*O*-MᴇNPOC-2'-ᴅᴇᴏxʏᴛʜʏᴍɪᴅɪɴᴇ-5'-*O*-(2-ᴄʏᴀɴᴏᴇᴛʜʏʟ)-*N,N*-ᴅɪɪꜱᴏᴘʀᴏᴘʏʟᴘʜᴏꜱᴘʜᴏʀᴀᴍɪᴅɪᴛᴇ (3'-MᴇNPOC-ᴅT-CEP)

^{1}H NMR (CDCl$_3$, 400 MHz): δ 8.80–8.58 (br m, 1H, H_{N3}), 7.64, 7.53 (2s, 1H, H_{C6}), 7.50 (s, 1H, $H_{Ar-MeNPOC}$), 7.07 (s, 1H, $H_{Ar-MeNPOC}$), 6.45–6.38 (m, 1H, $H_{C1'}$), 6.37–6.22 (m, 1H, $H_{CH-MeNPOC}$), 6.14 (s, 2H, $H_{OCH2O-MeNPOC}$), 5.25–5.12 (br m, 1H, $H_{C3'}$), 4.24, 4.27, 4.18, 4.15 (4 s, 1H, $H_{C4'}$), 3.97–3.73 (br m, 4H, $H_{C5', CE-a}$), 3.66–3.52 (br m, 2H, H_{CH-iPr}), 2.67–2.60 (m, 2H, H_{CE-b}), 2.55–2.34 (m, 1H, $H_{C2'a}$), 2.24–2.07 (br m, 1H, $H_{C2'b}$), 1.94(s, 3H, H_{CH3-C5}), 1.71–1.52 (2s, 3H, $H_{CH3-MeNPOC}$), 1.30–1.07 (mm, 12H, $H_{CH3-iPr}$) ^{31}P NMR (CDCl$_3$, 162 MHz): δ +149.88, 149.24, 149.14.

3.2. Substrate Preparation

Glass microscope slides were cleaned by soaking successively in peroxysulfuric acid (Nanostrip) for 15 min, followed by 10% aqueous NaOH at 70°C for 3 min, and then 1% aqueous HCl (1 min), rinsing thoroughly with deionized water between each step, and then spin dried for 5 min under a stream of nitrogen at 35°C. The slides were then silanated for 15 min in a gently agitating 1% solution of *N,N-bis*(2-hydroxyethyl)-3-aminopropyltriethoxysilane in 95:5 ethanol:water, rinsed thoroughly with 2-propanol, and then deionized water, and finally spin dried for 5 min at 90–110°C.

3.3. Photolithographic Array Synthesis

Array experiments were performed on an Affymetrix Classic Array Synthesizer, which consisted of a custom-built automated exposure system and a flowcell (1-in. diameter, 0.03-in. depth) linked to a modified Applied

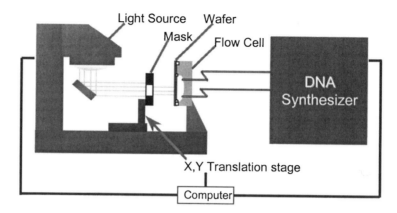

Fig. 4. Photolithographic oligonucleotide array synthesizer.

Biosystems model 392 DNA synthesizer (**Fig. 4**). Substrates were secured to the flowcell by a vacuum grip on the flowcell block. Phosphoramidites were used at a concentration of 50 mM in dry acetonitrile, and standard ancillary reagents were used for the coupling, capping, and oxidation steps. Reagent delivery from the synthesizer was controlled using OligoNet™ software, with minor adjustments made to the standard coupling protocol to accommodate the particular volume and mixing requirements of the flowcell. In cycles adapted for photolithographic synthesis, the detritylation step was replaced with a relay closure followed by a pause to allow the exposure system to perform the auto-mated mask positioning/alignment and timed light exposure. In "couple-only" cycles, the deprotection step was omitted entirely. Exposures were made through a 2 OD chrome-on-quartz mask in contact with the back of the sub-strate while the substrate remained clamped to the flowcell. Light was pro-jected horizontally from a 500-W collimated light source equipped with Ushio model ush508sa super high-pressure mercury lamp and dichroic reflectors to provide output in the near-UV spectral range (≥340 nm). Light intensity (mW/cm^2) was measured through the substrate and mask, with an HTG model 100B power meter equipped with a 365(±20)-nm bandpass filter. Prior to the exposure step, the flowcell may be filled with a solvent, such as dioxane, deliv-ered automatically by the DNA synthesizer (bottle no. 10), or dried thoroughly with argon. **Table 1** summarizes the steps of a typical photolithographic syn-thesis cycle.

3.4. Surface Fluorescence Staining

Typically, the final step of a photolysis or synthesis efficiency experiment (**Subheadings 3.7.1.** and **3.7.2.**) is fluorescence staining, wherein a fluorescein

Table 1
Synthesis Cycle for Photolithographic Synthesis

Step	Reagent	Repetitions	Duration
Wash	ACN	2	10 s
Dry	Argon	1	15 s
Couple	TET + amidite	2	25 s
Wash	ACN	1	10 s
Cap	AA/NMI/LUT/THF	3	10 s
Wash	ACN	1	10 s
Oxidize	I$_2$/H$_2$O/PYR/THF	3	10 s
Wash	ACN	4	10 s
Dry	Argon	1	25 s
Add solvent	*Solvent (optional)*	*1*	*12 s*
Exposure	Activate UV source, wait	1	$8 - 10 \times T_{1/2}$
Wash	ACN	2	10 s
Dry	Argon	1	15 s

Abbreviations: ACN, acetonitrile; TET, 0.45 M tetrazole in ACN; AA, acetic anhydride; NMI, *N*-methylimidazole; LUT, 2.6-lutidine; THF, tetrahydrofuran; PYR, pyridine.

phosphoramidite (Fluoreprime™, 5 mM, in a solution containing 50 mM DMT-T-CEP in acetonitrile) is coupled to the free hydroxyl groups on the substrate using the standard coupling protocol. The fluorescein phosphoramidite is diluted with a nonfluorescent phosphoramidite to avoid quenching interactions between fluorophores at high surface densities *(4)*.

3.5. Deprotection of the Array

Deprotection of DNA probe arrays in the usual aqueous ammonia solutions will lead to substantial and uncontrollable probe loss as a result of hydrolytic cleavage of the surface organosilane bonded phase. This can be avoided by using organic amine deprotecting agents in a nonaqueous solvent such as ethanol or acetonitrile. Complete deprotection of oligonucleotide arrays can be achieved by incubating the substrate in a solution of 1,2-diaminoethane in ethanol (50% [v]) for 4 h at room temperature, followed by rinsing with cold deionized water, and then drying under a stream of nitrogen.

3.6. Analysis of Array Synthesis

3.6.1. Surface Photolysis Rates

Figure 5 illustrates a typical strategy for determining the rates of photolytic deprotection for photolabile monomers attached to the support. In these

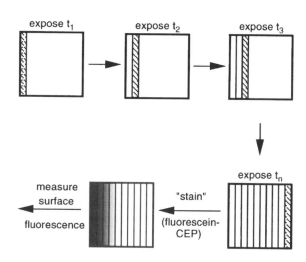

Fig. 5. Masking/synthesis scheme for determining surface photolysis rates. (Reprinted with permission from *J. Am. Chem. Soc.* **119,** 5081–5090. Copyright 1997, American Chemical Society.)

experiments, the silanated substrates were first modified with an HEG linker by coupling $DMT-[OCH_2CH_2]_6O-CEP$, capping the support with AmCAP using a standard couple-only cycle, detritylating the linker, and then coupling the desired photolabile monomer to the linker. A photolithographic mask is positioned over the back of the substrate to allow a selected portion of the substrate, in this case an open vertical aperture 0.8×12.8 mm, to be photolysed by exposure to near-UV light from a constant-intensity mercury light source for a predetermined duration. The mask is then translated horizontally 0.8 mm, placing the aperture of the mask over an adjacent region for a subsequent, longer exposure. This process is continued, increasing the length of exposure each time, to generate an array of stripes across the chip with a gradient of increasing exposure dose, and therefore increasing the extent of deprotection. The pattern of surface deprotection is then "stained" by coupling the fluorescein phosphoramidite mixture.

In a final step, the isobutyryl protecting groups are removed from the bound fluorescein by immersing the substrate in 1,2-diaminoethane in ethanol for 1 h. The resulting surface fluorescence image, acquired with a scanning confocal microscope, is shown in **Fig. 6,** and the fluorescence intensity change vs exposure time, extracted from the image, is plotted in **Fig. 7.** The change in surface fluorescence with exposure time followed a first-order exponential increase from which the photolysis rate constant or half-life ($t_{1/2}$, was obtained by simple curve-fitting analysis.

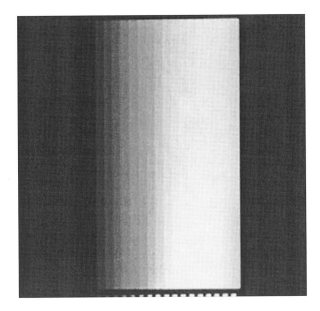

Fig. 6. Surface fluorescence image of a substrate prepared as described in **Fig. 5**. (Reprinted with permission from *J. Am. Chem. Soc.* **119,** 5081–5090. Copyright 1997, American Chemical Society.)

3.6.2. Efficiency of Stepwise Synthesis

Figure 8 illustrates the methodology used to estimate stepwise synthesis efficiencies for synthesis with photolabile monomers. In this experiment, the silanated substrates were first modified with an HEG linker by coupling MeNPOC-$[OCH_2CH_2]_6O$-CEP and capping with AmCAP using a standard couple-only cycle. A mask with a rectangular aperture was positioned over the substrate and exposed for a sufficient time to allow complete photolysis (10 half-lives) of the linker. An MeNPOC-nucleoside phosphoramidite was coupled to the exposed region of the support, and after capping and oxidation, the mask was offset horizontally by $1/n \times W$, where W is the width of the open reticle and n is the oligomer length to be tested. The photolysis was repeated, and a second monomer was added. This process was repeated for a total of n cycles of photolysis and coupling to generate a duplicate set of stripes on the support comprising a complete set of oligomers of length 1 to n. On completion, half of the array was subjected to a final full-field photolysis, as shown, to release photolabile groups from the 5' termini of the completed oligomers, including a region of the previously unphotolysed linker for comparison ($n = 0$ control). Fluorescein-CEP "stain" was then added to label the free 5'-hydroxyl

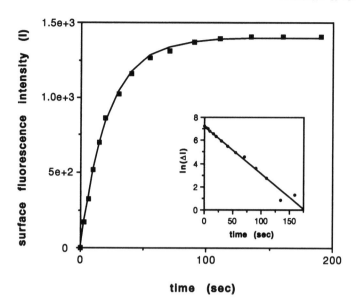

Fig. 7. Plot of fluorescence intensity vs exposure time, obtained from the image shown in **Fig. 6**. Intensity values were taken as the average pixel intensity in counts per second, in each region of the substrate exposed for a given time. The data have been fit to a first-order exponential (solid line). Inset shows linear plot of $\ln(\Delta I = I_\infty - I_t)$ vs time. (Reprinted with permission from *J. Am. Chem. Soc.* **119**, 5081–5090. Copyright 1997 American Chemical Society.)

groups for quantitation as described above. The unphotolysed half of the array was left to provide an internal control for background fluorescence owing to nonspecific binding of the fluorescein in regions of the substrate where synthesis had occurred.

Figure 9 shows the fluorescence image of a $(dC)_{0\text{-}12\text{-}0}$ array synthesized with 5'-MeNPOC phosphoramidites. Fluorescence decreases toward the center of the pattern since the yield of the full-length oligomer decreases with increasing length. After correcting for background fluorescence, the relative yield for each step in the synthesis was calculated from the ratio of the fluorescence intensities in adjacent stripes, as in **Eq. 1**:

$$\% \text{ yield (step } n) = 100 \times (I_n/I_{n-1}) \tag{1}$$

in which I_n is the fluorescence intensity for oligomer of length $= n$.

3.6.3. Experimental Arrays for Hybridization

For experimental purposes, a variety of simple arrays can be prepared to allow rapid comparisons of the performance of various synthesis or surface

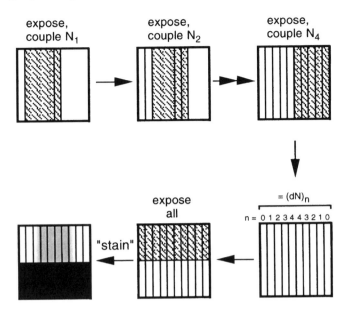

Fig. 8. Masking/synthesis scheme for determining surface stepwise synthesis efficiency. (Reprinted with permission from *J. Am. Chem. Soc.* **119,** 5081–5090. Copyright 1997, American Chemical Society.)

chemistries and process protocols. In some instances, it may be sufficient to synthesize a single test sequence in a "checkerboard" pattern, such as when relative hybridization rates or intensities are being compared. Another example is a small, 256-probe array such as the one illustrated in **Fig. 10**. This array is made in 22 synthesis steps according to the sequence of masking and base addition outlined in **Fig. 11**, and was used to compare the relative hybridization characteristics of 3'–5', vs 5'–3 direction synthesis chemistry, and of standard DNA probes vs "modified" probes containing 2'-*O*-methyl and 2,6-diaminopurine.

3.6.4. Hybridization to Test Arrays

Typical array hybridizations were carried out in a flowcell with the labeled oligonucleotide at 10 nM concentration in 6X SSPE buffer for 0.5–2 h at 22–45°C. After removal of the oligonucleotide solution, the array was washed with 6X SSPE buffer, and then imaged on a fluorescence scanner (*see* **Subheading 3.6.5.**).

3.6.5. Surface Fluorescence Imaging

The pattern and intensity of surface fluorescence on stained or hybridized arrays was imaged with a specially constructed scanning laser-induced fluo-

Fig. 9. Surface fluorescence image of an array of poly(dC) oligomers ranging in length from 1–12, prepared as outlined in **Fig. 8**. (Reprinted with permission from *J. Am. Chem. Soc.* **119**, 5081–5090. Copyright 1997, American Chemical Society.)

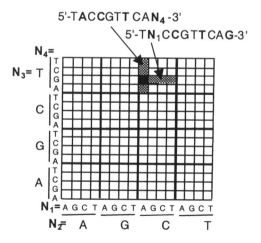

Fig. 10. Test array of 10-mer sequence containing four variable base positions.

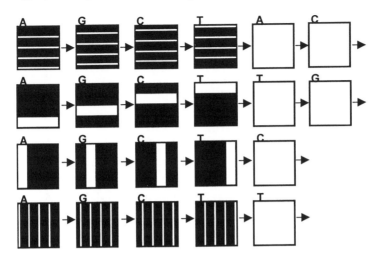

Fig. 11. Sequence of masking/synthesis steps for 10-mer test array shown in **Fig. 10.**

rescence confocal microscope, which employed excitation with a 488-nm argon ion laser beam focused to a 3-μ spot at the substrate surface. Emitted light was collected through confocal optics with a 530(±15)-nm bandpass filter and detected with a photomultiplier tube equipped with photon-counting electronics. Output intensity values (photon counts/s) are proportional to the amount of surface-bound fluorescein, so that relative amounts of bound fluorescein molecules within different regions of the substrate could be determined by direct comparison of the observed surface fluorescence intensities. Intensity values were corrected for nonspecific background fluorescence, taken as the surface fluorescence within the nonilluminated regions of the substrate.

4. Discussion

Photolabile MeNPOC-protected nucleoside phosphoramidite monomers can be readily prepared from the appropriate base-protected nucleosides (**Fig. 3**). In the case of 2'-deoxynucleosides, the reaction with MeNPOC-Cl in pyridine shows good selectivity for the 5'-OH. However, 2'-*O*-methylribonucleosides give intractable mixtures of 3' and 5' carbonylation products, and for these, transient 3'-*O*-silylation provided a more expedient route to the 5'-MeNPOC nucleoside. The optimum choice for the base exocyclic amine protecting groups are those that can be rapidly removed in ethylenediamine, without the formation of side products on cytidine. Deprotection of DNA probe arrays in the usual aqueous ammonia solutions leads to substantial and uncontrollable probe loss from the substrate owing to hydrolytic cleavage of the organosilane surface-bonded phase. This can be avoided by using organic amine deprotecting agents in a nonaqueous solvent such as ethanol or acetonitrile. Complete deprotection of oligonucleotide arrays is achieved by incubating the substrate in a 50% solution of ethylenediamine in ethanol for 4 h at room temperature. Another important factor to consider is that the base protecting groups can influence the photochemical synthesis yields *(4)*. Thus, for both C and G, isobutyryl is the preferred protecting group, and for A, phenoxyacetyl.

The surface-fluorescence-based testing protocols described herein provide a rapid means of evaluating photolysis rates and stepwise photochemical synthesis yields *(4,8)*. These are critical parameters to monitor when synthesizing oligonucleotide arrays, in order to maximize and maintain consistency in probe synthesis yields, and therefore array performance. Photolysis rates for removal of any photoremovable group will be dependent on the photochemical action spectrum of the protecting group, the energy spectrum and intensity of the exposure system, and the presence of solvents or coreactants. In the case of certain photolabile groups, photocleavage rates may also show a dependence on the nucleic acid base with which it is associated, or the length of the oligonucleotide chain *(6,8)*. Photolysis rates for MeNPOC nucleotides do not depend on either of these factors *(4)*, and this is a matter of some convenience because it allows the use of a single exposure setting for the entire array synthesis process ($0.5 \ J/t_{1/2}$, based on light intensities measured at 365 nm from a filtered Hg source). Complete photolysis is attained after exposure for 8–10 half-lives.

The other chemical reactions involved in the base addition cycles (coupling, capping, and oxidation) use reagents in a vast excess over surface synthesis sites, and provided that sufficient time is allowed for completion, they will be essentially quantitative. This is readily confirmed by carrying out probe synthesis using DMT monomers with chemical (TCA) deprotection on the array synthesizer. In this case, the process consistently shows stepwise yields in excess of 98% *(4)*. The net yields of the photolysis reaction are therefore the primary factor on which stepwise synthesis efficiencies depend, and, typically, these yields are observed to be subquantitative *(4,8,11)*. For light-directed synthesis using MeNPOC chemistry, the average stepwise efficiency of oligonucleotide synthesis is in the range of 85–95% (**Table 2**). Interestingly, the photocleavage yields depend on the nucleotide base, whereas the photolysis rates do not.

Apart from the target labeling chemistry and the instrumentation used to detect binding to the array, hybridization thermodynamics is the major determinant of the array's performance in terms of sensitivity and specificity. Thus, array performance is a complex function of many parameters including probe length, sequence, base composition, and surface density; the characteristics of the linker groups through which probes are attached to the support; the target sample concentration, complexity, fragment length, and secondary structure;

Table 2
Average Stepwise Yields for Synthesis of (N)$_{12}$
Homopolynucleotides Using MeNPOC-N-Phosphoramidites

Entry no.	Nucleoside (N)	MeNPOC site	Average stepwise yield[a]
1	dA(pac)	5'	95
2	dG(ibu)	5'	92
3	dG(pac)	5'	86
4	dC(ibu)	5'	96
5	T	5'	94
6	dI	5'	90
7	dD(pac)$_2$	5'	85
8	2'-OMe-rD(pac)$_2$	5'	80
9	2'-OMe-rA(bz)	5'	85
10	2'-OMe-^{5Me}U	5'	90
11	T	3'	91
12	dC(ibu)	3'	94
13	dG(ibu)	3'	85
14	dA(bz)	3'	93

[a]Photodeprotection step carried out in the presence of dioxane under near-UV output from Hg source (365 nm/[27.5 mW·cm^2]).

and the hybridization reaction conditions (e.g., [salt], pH, temperature, time). These parameters must all be taken into consideration to build accurate, reproducible hybridization-based assay protocols with DNA probe arrays.

For certain applications, nonstandard probe chemistries may be used to impart specific properties to an array. For example, "reverse" (5–3') synthesis enables probes to be linked to the support at the 5' end, leaving the 3' end available for enzymatic extension or ligation reactions *(18)*. "Reverse" arrays can readily be synthesized using 3'-MeNPOC 5'-phosphoramidite monomers, and these arrays have hybridization characteristics essentially equivalent to standard 3'–5' arrays, except for small, predictable differences in their response to mismatches near the ends of the probes.

Another way that nonstandard chemistries can be exploited is to enhance target binding to probe sequences, which under normal conditions, form less stable duplexes. Oligonucleotide arrays containing many thousands of probe sequences encompass a fairly broad range of hybridization thermodynamics, and a corresponding range of signal intensities is observed when they are hybridized with labeled target sequences under a given set of experimental conditions. A/T-rich probe sequences frequently display lower hybridization signal intensities, compared with sequences with high-G/C content, because they generally form less stable duplexes. This is a particular concern when

A   **B**

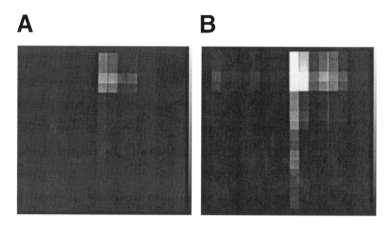

Fig. 12. Enhancement of hybridization with analog probes. The 256-decanucleotide test arrays outlined in **Fig. 10** were hybridized with an RNA oligo target [5'-fluorescein-rCTGAACGGTA-3'] = 10 n*M*, in 6X SSPE buffer for 1 h at 35°C, then washed with 6X SSPE at 22°C. (**A**) Standard A/T probe array. (**B**) Analog D_{OMe}/T_{OMe} probe array. The observed brightness at any particular cell is proportional to the amount of labeled target hybridized to the probe at that site.

very stringent hybridization conditions are required to break up secondary structure in certain RNA targets, or to increase the discrimination of arrays when used for the analysis of high-complexity mRNA samples *(19,20)*. The use of TMACl has been explored as a means of improving hybridization to A/T-rich probe sequences, but this has limited practical use and the disadvantage of being toxic and corrosive *(21)*. We have found that the sensitivity of A/T-rich probe sequences in arrays can be dramatically improved by the introduction of certain nucleotide analogs that are capable of increasing overall duplex stability while maintaining a high degree of base specificity.

We have studied the analogs 2'-deoxy-2,6-diaminopurine (dD) and the 2'-*O*-methylribonucleotides 2'-*O*-Me-A, 2'-OMe-D as replacements for deoxyadenosine; and 2'-OMe-5-Me-U as a replacement for thymidine. **Figure 12** provides a simple illustration of how the construction of probes with these building blocks can significantly increase hybridization signal intensities with RNA targets, even in small experimental arrays of 10-mer sequences, such as the one shown in **Fig. 10**. Replacement of A with dD has been suggested as a means of enhancing the stability of A-rich sequences *(22–29)*; however, the stabilizing effect is not entirely uniform over all sequences. The introduction of dD in the 10-mer array has a minor impact on hybridization of a complementary oligonucleotide, 5'-fluorescein-rCTGAACGGTA-3' (**Fig. 12A,B**). It does, however, consistently stabilize duplexes between RNA targets and probes with

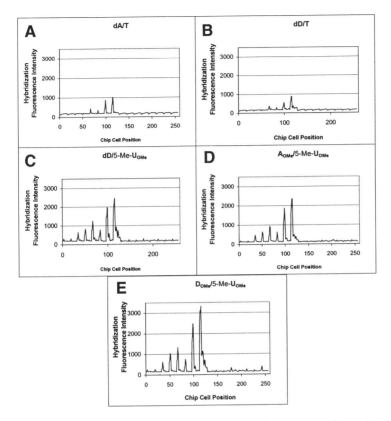

Fig. 13. Histogram plots of observed hybridization fluorescence intensity for 256-decanucleotide test arrays that were hybridized with RNA oligo target as described in **Fig. 12**. The arrays were synthesized with various combinations of the A and T analogs dD, 2'-O-Me-A, 2'-OMe-D and 2'-OMe-5-Me-U as indicated on each plot.

poly-dA tracts *(28,29)*, and this has been observed consistently in probe array experiments. For example, the T_{10} and U_{10} targets both hybridize with >40 times higher affinity to $(dD)_{10}$ probe than to $(dA)_{10}$ probe on an array, whereas an alternating copolymer $(AU)_5$ target hybridizes only approx 5 times better to $(TdD)_5$ than to $(TdA)_{10}$ probes, and no enhancement was observed with $(dAT)_5$ target.

Substantial overall improvement in RNA hybridization intensities is observed when dA and T in the probe are replaced with 2'-OMe-D, 2'-OMe-5-Me-U analogs. Because of their stabilizing effect on duplex formation with complementary RNA target sequences *(30)*, 2'-*O*-alkyl-modified oligonucleotides have been considered extensively as antisense therapeutic agents and more recently as improved probes for RNA detection *(31)*. **Figure 13C–E**

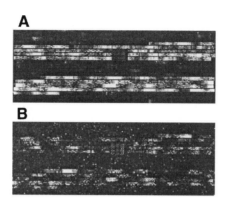

Fig. 14. Analogs improve the sensitivity of hybridization of A/T-rich RNA targets to an HIV PRT 440 resequencing GeneChip probe array at 50°C in 6X SSPE buffer. (**A**) 20-mer dD/$^{5\text{-Me}}$U$_{\text{OMe}}$-substituted probe array. (**B**) Standard 20-mer A/T probe array. Scan images are normalized to the same fluorescence intensity scale. In this experiment, the analogs increased the "base-call" accuracy of the array to 98.6%, from 88.6% (standard A/T array). Target sample preparation, labeling, and hybridizations were carried out according to published protocols *(33,34)*. Reprinted from **ref. 2**, p. 1294, courtesy of Marcel Dekker, Inc.

shows how substitution of dA and T in the 10-mer probe array with 2'-OMe-D, 2'-OMe-5-Me-U analogs increases relative hybridization intensities two- to threefold, even when the substitution is nonuniform (i.e., Gs and Cs remain unmodified). Interestingly, substituting dA in the probes with 2'-*O*-Me-A consistently shows little, if any, impact on hybridization signal intensities.

In some array designs, the improvement afforded by these duplex-stabilizing factors can significantly improve the performance of arrays under conditions of high stringency (**Fig. 14**). They can also allow a substantial reduction in the length of probes (e.g., 12- to 14-mers) needed to obtain performance characteristics that were previously only achievable with arrays of longer (e.g., 20-mer) probes. This has been recently demonstrated in high-density GeneChip probe arrays designed for analyzing drug resistance in human immunodeficiency virus (HIV) *(32)*. Improving the hybridization performance of arrays of short probes will also be important for the development of generic probe arrays based on sets of all-*n*-mer sequence *(33)*.

Acknowledgments

Portions of **Subheadings 2.** and **3.** were reproduced with permission from *J. Am. Chem. Soc.* **119,** 5081–5090 (Copyright 1997, American Chemical

Society). We wish to thank the following individuals for their assistance in the preparation of MeNPOC phosphoramidite reagents: Dale Barone (Affymetrix), Nam Ngo (present address: CTGen, Milpitas, CA), and Mortezai Vaghefi (Trilink Biotechnologies, San Diego, CA).

References

1. (1999) The chipping forecast. *Nat. Genet.* **21(1Suppl.),** 1–60.
2. Fodor, S. P. A., Read, J. L., Pirrung, M. C., Stryer, L. T., Lu, A., and Solas, D. (1991) Light-directed, spatially addressable parallel chemical synthesis. *Science* **251,** 767–773.
3. Pease, A. C., Solas, D., Sullivan, E. J., Cronin, M. T., Holmes, C. P., and Fodor, S. P. A. (1994) Light-generated oligonucleotide arrays for rapid DNA sequence analysis. *Proc. Natl. Acad. Sci. USA* **91,** 5022–5026.
4. McGall, G. H., Barone, D., Diggelmann, M., Fodor, S. P. A., Gentalen, E., and Ngo, N. (1997) The efficiency of light-directed synthesis of DNA arrays on glass substrates. *J. Am. Chem. Soc.* **119,** 5081–5090.
5. Lipshutz, R., Fodor, S. P. A., Gingeras, T. R., and Lockhart, D. J. (1999) High-density synthetic oligonucleotide arrays. *Nat. Genet.* **21(1Suppl.),** 20–24.
6. Pirrung, M. C. and Bradley, J.-C. (1995) Dimethoxy benzoin carbonates: photo-chemically-removable alcohol protecting groups suitable for phosphoramidite-based DNA synthesis. *J. Org. Chem.* **60,** 1116–1117.
7. Pirrung, M. C. and Bradley, J.-C. (1995) Comparison of methods for photochemical phosphoramidite-based DNA synthesis. *J. Org. Chem.* **60,** 6270–6276.
8. Pirrung, M. C., Fallon, L., and McGall, G. H. (1998) Proofing of photolithographic DNA synthesis with 3',5'-dimethoxybenzoinyloxy-carbonyl-protected deoxy-nucleoside phosphoramidites. *J. Org. Chem.* **63,** 241–246.
9. McGall, G. H. (1997) The fabrication of high density oligonucleotide arrays for hybridization-based sequence analysis, in *Biochip Arrays and Integrated Devices for Clinical Diagnostics* (Hori, W., ed.), IBC Library Series, Southboro, MA, pp. 2.1–2.33.
10. Hasan, A., Stengele, K.-P., Giegrich, H., Cornwell, P., Isham, K.R., Sachleben, R., Pfleiderer, W., and Foote, R. S. (1997) Photolabile protecting groups for nucleosides: synthesis and photodeprotection rates. *Tetrahedron* **53,** 4247–4262.
11. Giegrich, H., Eisele-Buhler, S., Herman, C., Kvasyuk, E., Charubala, R., and Pfleiderer, W. (1998) New photolabile protecting groups in nucleoside and nucle-otide chemistry—synthesis, cleavage mechanisms and applications. *Nucleosides Nucleotides* **17,** 1987–1996.
12. McGall, G. H., Labadie, J., Brock, P., Wallraff, G., Nguyen, T., and Hinsberg, W. (1996) Light-directed synthesis of high-density oligonucleotide arrays using semi-conductor photoresists. *Proc. Natl. Acad. Sci. USA* **93,** 13,555–13,560.
13. Wallraff, G., Labadie, J., Brock, P., DiPietro, R., Nguyen, T., Huynh, T., Hinsberg, W., and McGall, G. (1997) DNA chips. *Chemtech* **27,** 22–32.

14. Beecher, J. E., McGall, G. H., and Goldberg, M. J. (1997) Chemically amplified photolithography for the fabrication of high density oligonucleotide arrays. *Am. Chem. Soc. Div. Polym. Mater. Sci. Eng.* **76,** 597–598.

15. van Boom, J. H. and Wreesmann, C. J. T. (1984) Chemical synthesis of small oligoribonucleotides in solution, in *Oligonucleotide Synthesis: A Practical Approach*, (Gait, M. J., ed.), Oxford University Press, New York, NY, pp. 153–183.

16. Pirrung, M. C., Shuey, S. W., Lever, D. C., and Fallon, L. (1994) A convenient procedure for the deprotection of silylated nucleosides and nucleotides using tri-ethylamine trihydrofluoride. *Bioorgan. Med. Chem. Lett.* **4,** 1345–1346.

17. Barone, A. D., Tang, A.-Y., and Caruthers, M. H. (1984) *In situ* Activation of bis-dialkylaminophosphines—a new method for synthesizing deoxyoligonucleotides on polymer supports. *Nucleic Acids Res.* **12,** 4051–4061.

18. Broude, N. E., Sano, T., Smith, C. L., and Cantor, C. R. (1994) Enhanced DNA sequencing by hybridization. *Proc. Natl. Acad. Sci. USA* **91,** 3072–3076.

19. Chee, M. S., Huang, X., Yang, R., Hubbell, E., Berno, A., Stern, D., Winkler, J., Lockhart, D. J., Morris, M. S., and Fodor, S. P. A. (1996) Accessing genetic infor-mation with high-density DNA arrays. *Science* **274,** 610–614.

20. Lockhart, D. J., Dong., H., Byrne, M. C., Follettie, M. T., Gallo, M. V., Chee, M. S., Mittmann, M., Wang, C., Kobayashi, M., Horton, H., and Brown, E. L. (1996) Expression monitoring by hybridization to high-density oligonucleotide arrays. *Nat. Biotechnol.* **14,** 1675–1680.

21. Maskos, U. and Southern, E. M. (1993) A study of oligonucleotide reassociation using large arrays of oligonucleotides synthesized on a glass support. *Nucleic Acids Res.* **21,** 4663–4669.

22. Gaffney, B. L., Marky, L. A., and Jones, R. A. (1984) The influence of purine 2-amino group on DNA conformation and stability-II. *Tetrahedron* **40,** 3–13.

23. Chollet, A., Chollet-Damerius, A., and Kawashima, E. H. (1985) Synthesis of oligonucleotides containing the base 2-aminoadenine. *Chem. Scripta* **26,** 37–40.

24. Cheong, C., Tinoco, J., Jr., and Chollet, A. (1988) Quantitative measurements on the duples stability of 2,6-diaminopurine and 5-chlorouracil nucleotides using enzymatically synthesized oligomers. *Nucleic Acids Res.* **16,** 5115–5122.

25. Hoheisel, J. D. and Lehrach, H. (1990) Quantitative measurements on the duplex stability of 2,6-diaminopurine and 5-chloro-uracil nucleotides using enzymati-cally synthesized oligomers. *FEBS Lett.* **274,** 103–106.

26. Chollet, A., Chollet-Damerius, A., and Kawashima, E. H. (1994) Synthesis of oligonucleotides containing the base 2-aminoadenine. *Chem. Scripta* **26,** 37–40.

27. Prosnyak, M. I., Veselovskaya, S. I., Myasnikov, V. A., Efremova, E. J., Potapov, V. K., Limborska, S. A., and Sverdlov, E. D. (1994) Substitution of 2-aminoadenine and 5-methylcytosine for adenine and cytosine in hybridization probes increases the sensi-tivity of DNA fingerprinting. *Genomics* **21,** 490–494.

28. Howard, F. B. and Miles, H. T. (1984) 2-NH$_2$A:T helicies in the ribo- and deoxypolynucleotide series. Structural and energetic consequences of NH$_2$A sub-stitution. *Biochemistry* **23,** 6723–6732.

29. Gryaznov, S. and Schultz, R. G. (1994) Stabilization of DNA:DNA and DNA:RNA duplexes by substitution of 2'-deoxyadenosine with 2'-deoxy-2-aminoadenosine. *Tetrahedron Lett.* **35,** 2489–2492.

30. Inoue, H., Hayase, Y., Imura, A., Iwai, S., Miura, K., and Ohtsuka, E. (1987) Synthesis and hybridization studies on two complementary nona(2'-O-methyl)ribonucleotides. *Nucleic Acids Res.* **15,** 6131–6148.

31. Majlessi, M., Nelson, N. C., and Becker, M. M. (1998) Advantages of 2'-O-methyl oligoribonucleotide probes for detecting RNA targets. *Nucleic Acids Res.* **26,** 2224–2229.

32. Fidanza, J. A. and McGall, G. H. (1999) High-density nucleoside analog probe arrays for enhanced hybridization. *Nucleosides Nucleotides* **18,** 1293–1295.

33. Gunderson, K. L., Huang, X. C., Morris, M. S., Lipshutz, R. J., Lockhart, D. J., and Chee, M. S. (1998) Mutation detection by ligation to complete N-mer DNA arrays. *Genome Res.* **8,** 1142–1153.

34. Kozal, M. J., Shah, N., Shen, N., Yang, R., Fucini, R., Merigan, T., Richman, D., Morris, D., Hubbell, E., Chee, M. S., and Gingeras, T. G. (1996) Extensive polymorphisms observed in HIB-1 clade B protease gene using high-density oligonucleotide arrays. *Nat. Med.* **7,** 753–759.

6

Automated Genotyping Using the DNA MassArray™ Technology

Christian Jurinke, Dirk van den Boom, Charles R. Cantor, and Hubert Köster

1. Introduction
1.1. Markers Used for Genetic Analysis

The ongoing progress in establishing a reference sequence as part of the Human Genome Project *(1)* has revealed a new challenge: the large-scale identification and detection of intraspecies sequence variations, either between individuals or populations. The information drawn from those studies will lead to a detailed understanding of genetic and environmental contributions to the etiology of complex diseases.

The development of markers to detect intraspecies sequence variations has evolved from the use of restriction fragment length polymorphisms (RFLPs) to microsatellites (short tandem repeats [STRs]) and very recently to single nucleotide polymorphisms (SNPs).

Although RFLP markers *(2)* are useful in many applications, they are often of poor information content, and their analysis is cumbersome to automate. STR markers *(3)*, by contrast, are fairly highly informative (through their highly polymorphic number of repeats) and easy to prepare using polymerase chain reaction (PCR)-based assays with a considerable potential for automation. However, using conventional gel electrophoresis-based analysis, typing of large numbers of individuals for hundreds of markers still remains a challenging task.

Within the last few years, much attention has been paid to discovery and typing (scoring) of SNPs and their use for gene tracking *(4,5)*. SNPs are biallelic single-base variations, occurring with a frequency of at least 1 SNP/1000 bp within the 3 billion bp of the human genome. Recently, a study on the sequence

From: *Methods in Molecular Biology, vol. 170: DNA Arrays: Methods and Protocols*
Edited by: J. B. Rampal © Humana Press Inc., Totowa, NJ

diversity in the human lipoprotein lipase gene suggested that the frequency of SNPs might be much higher *(6)*. The diversity in plant DNA, which would be relevant for agricultural applications, is five to seven times larger than in human DNA *(7)*.

Even though the use of SNPs as genetic markers seems to share the same limitations as relatively uninformative RFLPs, when used with modern scoring technologies, SNPs exhibit several advantages. Most interesting for gene tracking is that SNPs exist in the direct neighborhood of genes and also within genes. Roughly 200,000 SNPs are expected *(4)* in protein coding regions (so-called cSNPs) of the human genome. Furthermore, SNPs occur much more frequently than STRs and offer superior potential for automated assays.

1.2. Demand for Industrial Genomics

1.2.1. Genetics

The efforts of many researchers are dedicated to the exploration of the genetic bases of complex inherited diseases or disease predispositions. Studies are performed to identify candidate or target genes that may confer a predisposition for a certain disease *(8)*. Linkage analysis can be done as a genome wide screening of families; association or linkage disequilibrium analysis can be done with populations. Either approach can use STR or SNP markers. Once a potential candidate gene is discovered, a particular set of markers is compared between affected and unaffected individuals to try to identify functional allelic variations. To understand the genotype-to-phenotype correlation of complex diseases, several hundred markers need to be compared among several hundred individuals *(9,10)*. To get an impression of the complexity of the data produced in such projects, imagine a certain multifactorial disease in which predisposition is linked to, e.g., 12 genes. Consider that each of those 12 genes can be present in just two different alleles. The resulting number of possible genotypes (2 homozygotes and 1 heterozygote = 3 for each gene) is $3^{12} = 531,441$.

The whole process of drug development, including hunting for new target genes and especially the subsequent validation (significant link to a certain disease), will benefit from high-throughput, high-accuracy genomic analysis methods. Validated target genes can also be used for a more rational drug development in combination with genetic profiling of study populations during clinical trials.

1.2.2. Pharmacogenetics

Traits within populations, such as the ABO blood groups, are phenotypic expressions of genetic polymorphism. This is also the case for variations in response to drug therapy. When taken by poor metabolizers, some drugs cause exaggerated pharmacological response and adverse drug reactions. For example, tricyclic antidepressants exhibit order of magnitude differences in blood concentrations depending on the enzyme status of patients *(11)*. Pharmacogenetics is the study of genetic polymorphism in drug metabolism. Today, pharmaceutical companies screen individuals for specific genetic polymorphisms before entry into clinical trials to ensure that the study population is both relevant and representative. Targets for such screenings are cytochrome P450 enzymes or *N*-acetyltransferase isoenzymes (NAT1 and NAT2). Potential drug candidates affected by polymorphic metabolism include antidepressants, antipsychotics, and cardiovascular drugs.

1.2.3. Current Technologies

In addition to candidate gene validation and pharmacogenetics, many other applications such as clinical diagnostics, forensics, as well as the human sequence diversity program *(12)* are dealing with SNP scoring. In agricultural approaches, quantitative trait loci can be explored, resulting in significant breeding advances. Methods are required that provide high-throughput, parallel sample processing; flexibility; accuracy; and cost-effectiveness to match the different needs and sample volumes of such efforts.

Large-scale hybridization assays performed on microarrays have enabled relatively high-throughput profiling of gene expression patterns *(13)*. However, several issues must be considered in attempting to adapt this approach for the large-scale genotyping of populations of several hundred individuals. Hybridization chips for SNP scoring can potentially analyze in parallel several hundred SNPs per chip—with DNA from one individual. Therefore, several hundred hybridization chips would be needed for projects with larger populations. If during the course of a study an assay needs to be modified or new assays have to be added, all chips might have to be completely remanufactured.

Also, note that DNA hybridization lacks 100% specificity. Therefore, highly redundant assays have to be performed, providing a statistical result with a false-negative error rate of up to 10% for heterozygotes *(14)*. Finally, because of the inherent properties of repeated sequences, hybridization approaches are hardly applicable to STR analysis.

1.3. DNA MassArray Technology

Within the last decade, mass spectrometry (MS) has been developed to a powerful tool no longer restricted to the analysis of small compounds (some hundred Daltons) but also applicable to the analysis of large biomolecules (some hundred thousand Daltons). This improvement is mainly based on the invention of soft ionization techniques. A prominent example is matrix assisted

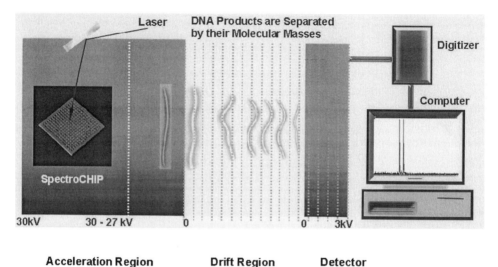

Fig. 1. Schematic drawing of the MALDI-TOF MS process, as used in the DNA MassArray method.

laser desorption/ionization (MALDI) time-of-flight (TOF) MS, developed in the late 1980s by Karas and Hillenkamp *(15)*.

The general principle of MS is to produce, separate, and detect gas-phase ions. Traditionally, thermal vaporization methods are used to transfer molecules into the gas phase. Most biomolecules, however, undergo decomposition under these conditions. Briefly, in MALDI MS, the sample is embedded in the crystalline structures of small organic compounds (called matrix), and the cocrystals are irradiated with a nanosecond ultraviolet-laser beam. Laser energy causes structural decomposition of the irradiated crystal and generates a particle cloud from which ions are extracted by an electric field. After acceleration, the ions drift through a field-free path (usually 1 m long) and finally reach the detector (e.g., a secondary electron multiplier) (*see* **Fig. 1**). Ion masses (mass-to-charge ratios, *m/z*) are typically calculated by measuring their TOF, which is longer for larger molecules than for smaller ones (provided their initial energies are identical). Because predominantly single-charged nonfragmented ions are generated, parent ion masses can easily be determined from the spectrum without the need for complex data processing and are accessible as numerical data for direct processing.

The quality of the spectra, which is reflected in terms of resolution, mass accuracy, and also sensitivity, is highly dependent on sample preparation and the choice of matrix compound. For this reason, the early applications of MALDI-TOF MS were mostly for analyzing peptides and proteins. The discovery of new matrix compounds for nucleic acid analysis *(16)* and the development of solid-phase sample conditioning formats *(17,18)* enabled the analysis of nucleic acid reaction products generated in ligase chain reaction or PCR *(19)*.

The more demanding DNA sequence determination with MALDI-TOF MS can be addressed using exonucleolytic digestion *(20)*, Sanger sequencing *(21)*, or solid-phase Sanger sequencing approaches *(22)*. These approaches are currently restricted to comparative sequencing, and the read length is limited to about 100 bases. Further improvements in reaction design and instrumentation *(23)* will surely lead to enhanced efficacy and longer read length. For genotyping applications, this limitation is not relevant because scientists at Sequenom (San Diego, CA) developed the primer oligo base extension (PROBE) reaction especially for the purpose of assessing genetic polymorphism by MS *(24)*. The PROBE assay format can be used for the analysis of deletion, insertion, or point mutations, and STR, and SNP analysis, and it allows the detection of compound heterozygotes. The PROBE process comprises a postPCR solid-phase primer extension reaction carried out in the presence of one or more dideoxynucleotides (ddNTPs) and generates allele-specific terminated extension fragments (*see* **Fig. 2**). In the case of SNP analysis, the PROBE primer binding site is placed adjacent to the polymorphic position. Depending on the nucleotide status of the SNP, a shorter or a longer extension product is generated. In the case of heterozygosity, both products are generated. After completion of the reaction, the products are denatured from the solid phase and analyzed by MALDI-TOF MS. In the example given in **Fig. 2**, the elongation products are expected to differ in mass by one nucleotide. **Figure 3** presents raw data for a heterozygous DNA sample analyzed by this PROBE assay. The two SNP alleles appear as two distinct mass signals. Careful assay design makes a high-level multiplexing of PROBE reactions possible.

In the case of STR analysis, a ddNTP composition is chosen that terminates the polymerase extension at the first nucleotide not present within the repeat *(25)*. For length determination of a CA repeat, a ddG or ddT termination mix is used. Even imperfect repeats harboring insertion or deletion mutations can be analyzed with this approach. **Figure 4** displays raw data from the analysis of a human STR marker in a heterozygous DNA sample. Both alleles differ by four CA repeats. The DNA polymerase slippage during amplification generates a pattern of "stutter fragments" (marked with an asterisk in **Fig. 4**). In the case of heterozygotes that differ in just one repeat, the smaller allele has higher intensities than the larger allele, because allelic and stutter signals are added together. A DNA MassArray compatible STR portfolio with a 5-cM intermarker distance is currently under development at Sequenom.

When compared to the analysis of hybridization events by detecting labels, even on arrays, the DNA MassArray approach differs significantly. The

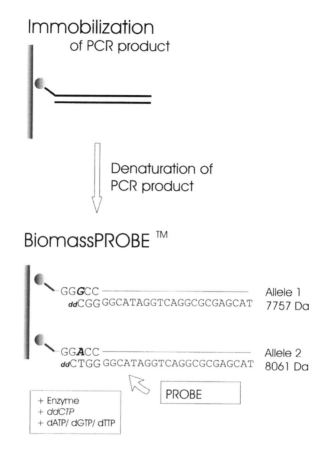

Fig. 2. Reaction scheme for the BiomassPROBE reaction.

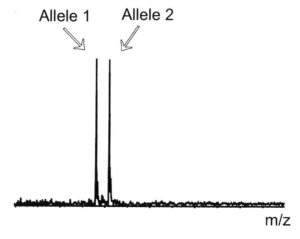

Fig. 3. Raw data of SNP analysis (heterozygous sample) using the BiomassPROBE reaction.

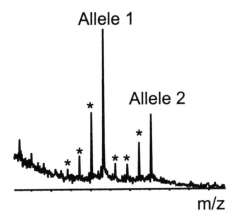

Fig. 4. Raw data of microsatellite analysis (heterozygous sample) using the BiomassPROBE reaction. Signals marked with an asterisk are stutter fragments (*see* **Subheading 1.3.**).

PROBE assay is designed to give only the relevant information. The mass spectrometric approach enables direct analyte detection with 100% specificity and needs no redundancy. This accuracy and efficacy is combined with sample miniaturization, bioinformatics, and chip-based technologies for parallel processing of numerous samples.

Now, the use of an advanced nanoliquid handling system based on piezoelectric pipets combined with surface-modified silicon chips permits an automated scanning of 96 samples in about 10 min. Currently, up to 10 SpectroCHIPs (960 samples) can be analyzed in one automated run using a Bruker/Sequenom SpectroSCAN mass spectrometer (*see* **Fig. 5**). The SpectroSCAN mass spectrometer addresses each position of the chip sequentially, collects the sum of 10 laser shots, processes and stores the data, and proceeds to the next spot of the chip. In **Fig. 6**, 96 raw data spectra from a heterozygous sample are depicted resulting from a SpectroCHIP with one sample spotted 96 times. Using a proprietary algorithm, masses as well as signal intensities are automatically analyzed and interpreted. After completion of analysis, the results are transferred to a database and stored as accessible genetic information (*see* **Fig. 7**). The database also provides a tool for visual control and comparison of spectra with theoretically expected results (*see* **Fig. 8**).

The DNA MassArray throughput in terms of genetic information output depends on the chosen scale. Using microtiter plates and 8-channel pipets, the analysis of 192 genotypes (two 96-well microtiter plates) a day is routine work. With the use of automated liquid handling stations, the throughput can be increased by a factor of about four. An automated process line was been devel-

Fig. 5. Sample holder for 10 SpectroCHIPs for use in the SpectroSCAN mass spectrometer.

oped during the last year to increase the throughput to an industrial scale. The automated process line integrates biochemical reactions including PCR setup, immobilization, PROBE reaction sample conditioning, and recovery from the solid-phase into a fully automated process with a throughput of about 10,000 samples per day.

2. Materials

2.1. PCR and PROBE Reaction

1. Dynabeads M-280 Streptavidin (Dynal, Oslo, Norway).
2. Separate PROBE stops mixes for ddA, ddC, ddG, and ddT (500 μM of the respective ddNTP and 500 μM of all dNTPs not present as dideoxynucleotides) (MassArray Kit; Sequenom).
3. 2X B/W buffer: 10 mM Tris-HCl, pH 7.5, 1 mM EDTA, 2 M NaCl (all components from Merck, Darmstadt, Germany).
4. 25% Aqueous NH$_4$OH (Merck, Darmstadt, Germany).
5. 10 mM Tris-HCl, pH 8.0 (Merck).
6. AmpliTaq Gold (Perkin-Elmer, Foster City, CA).
7. AmpliTaq FS (Perkin-Elmer).

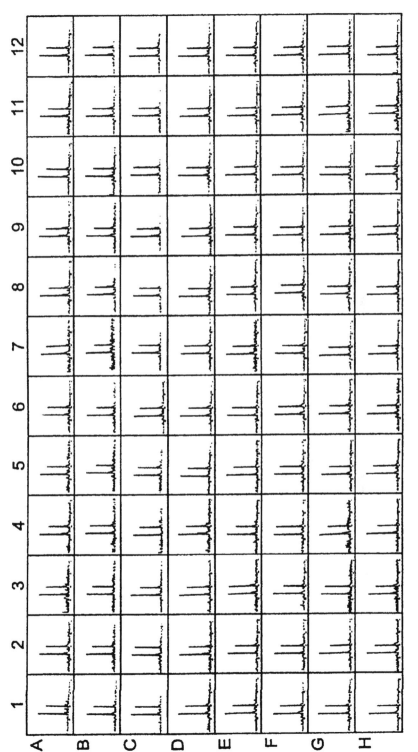

Fig. 6. Raw data generated during the analysis of one sample spotted 96 times on a SpectroCHIP.

Reaction_Details			Sample_Details					Assay_Details			
PlateNo.	PlateID	Well	SampleNo.	SampleID	PlateNo.	PlateID	Well	AssayNo.	Name	Result	Spectrum
1	11S7872	A1	1	14G88	1	23R902	A1	1	AMG	Male	11S7872_A1.sq
1	11S7872	A2	2	14G89	1	23R902	A2	1	AMG	Male	11S7872_A2.sq
1	11S7872	A3	3	14G90	1	23R902	A3	1	AMG	Female	11S7872_A3.sq
1	11S7872	A4	4	14G91	1	23R902	A4	1	AMG	Female	11S7872_A4.sq
1	11S7872	A5	5	14G92	1	23R902	A5	1	AMG	Male	11S7872_A5.sq
1	11S7872	A6	6	14G93	1	23R902	A6	1	AMG	Female	11S7872_A6.sq
1	11S7872	A7	7	14G94	1	23R902	A7	1	AMG	Male	11S7872_A7.sq
1	11S7872	A8	8	14G95	1	23R902	A8	1	AMG	Female	11S7872_A8.sq
1	11S7872	A9	9	14G96	1	23R902	A9	1	AMG	Male	11S7872_A9.sq
1	11S7872	A10	10	14G97	1	23R902	A10	1	AMG	Male	11S7872_A10.sq
1	11S7872	A11	11	14G98	1	23R902	A11	1	AMG	Female	11S7872_A11.sq
1	11S7872	A12	12	14G99	1	23R902	A12	1	AMG	Female	11S7872_A12.sq
1	11S7872	B1	13	14G100	1	23R902	B1	1	AMG	Male	11S7872_B1.sq
1	11S7872	B2	14	14G101	1	23R902	B2	1	AMG	Female	11S7872_B2.sq

Fig. 7. Sequenom data analysis software reports for automated sex typing using the DNA MassArray.

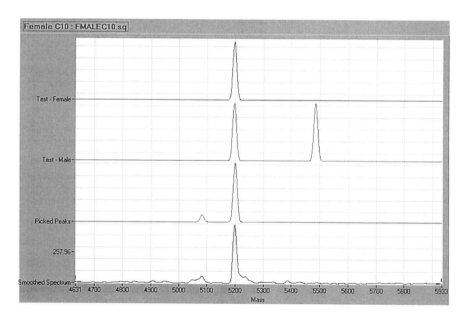

Fig. 8. Tool for visual comparison of spectra with the theoretical results.

8. Magnetic particle concentrator for microtiter plate or tubes (Dynal).
9. Specific PCR and PROBE primer (*see* **Note 1**).

2.2. Nanoliquid Handling and SpectroCHIPs

1. SpectroCHIP (Sequenom).
2. SpectroJET (Sequenom).

2.3. SpectroCHIP Analysis

1. SpectroSCAN (Sequenom).
2. SpectroTYPER (Sequenom).

3. Methods
3.1. PCR and PROBE Reaction

The following steps can be performed either in microtiter plates using multichannel manual pipettors or automated pipetting systems or on the single-tube scale.

3.1.1. Preparation of PCR

Perform one 50-µL PCR per PROBE reaction with 10 pmol of biotinylated primer and 25 pmol of nonbiotinylated primer (*see* **Note 2**).

3.1.2. Immobilization of Amplified Product

1. For each PCR use 15 µL of streptavidin Dynabeads (10 mg/mL).
2. Prewash the beads twice with 50 µL of 1X B/W buffer using the magnetic rack.
3. Resuspend the washed beads in 50 µL of 2X B/W buffer and add to 50 µL of PCR mix.
4. Incubate for 15 min at room temperature. Keep the beads resuspended by gentle rotation.

3.1.3. Denaturation of DNA Duplex

1. Remove the supernatant by magnetic separation.
2. Resuspend the beads in 50 µL of 100 mM NaOH (freshly prepared).
3. Incubate for 5 min at room temperature.
4. Remove and discard the NaOH supernatant by magnetic separation.
5. Wash three times with 50 µL of 10 mM Tris-HCl, pH 8.0.

3.1.4. PROBE Reaction

1. Remove the supernatant by magnetic separation, and add the following PROBE mix: 3 µL of 5X reaction buffer, 2 µL of PROBE nucleotide mix (ddA, ddC, ddG, or ddT with the respective dNTPs), 2 µL of PROBE primer (20 pmol), 7.6 µL of H$_2$O, 0.4 µL of enzyme (2.5 U).
2. The PROBE temperature profile comprises 1 min at 80°C, 3 min at 55°C, followed by 4 min at 72°C. Cool slowly to room temperature. Keep the beads resuspended by gentle rotation (*see* **Notes 3** and **4**).

3.1.5. Recovery of PROBE Products

1. After the reaction is completed, remove the supernatant by magnetic separation.
2. Wash the beads twice with 50 µL of 10 mM Tris-HCl, pH 8.0.
3. Resuspend the beads in 5 µL of 50 mM NH$_4$OH (freshly aliquoted from 25% stock solution).
4. Incubate for 4 min at 60°C.
5. Transfer the supernatant to a microtiter plate, and discard (or store) the beads.

3.2. SpectroCHIP Loading (see *Note 5*)

1. Fill containers with ultrapure water.
2. Initialize the nanoplotter.
3. Place the SpectroCHIP and microtiter plate on the nanoplotter (*see* **Note 6**).
4. Start the sample spotting program.

3.3. SpectroCHIP Scanning

1. Place the loaded SpectroCHIP on the sample holder.
2. Insert the sample holder into the SpectroSCAN.
3. Define which spots or chips have to be analyzed.
4. Choose analysis method and start the automated run.
5. Transfer the data to the processing server.

4. Notes

1. For PCR as well as PROBE primers it is useful to verify the masses before use. Primers that are not completely deprotected (mass shift to higher masses) or mixed with n-1 synthesis products should not be used.
2. Use asymmetric primer concentrations in PCR, with the nonbiotinylated primer in excess.
3. The length of the PROBE primer should not exceed 20–25 bases; try to have C or G at the 3' end, and avoid mismatches, especially at the 3' end.
4. The second temperature step in the PROBE program (55°C) depends on the primer length.
5. After the reaction, the beads can be stored for further reactions in Tris-HCl buffer at 4°C.
6. Be sure to handle SpectroCHIPs with gloves and avoid any contact with moisture.

References

1. Collins, F. S., Patrinos, A., Jordan, E., Chakravarti, A., Gesteland, R., Walters, L., and the members of DOE and NIH planning groups. (1998) New goals for the U.S. human genome project: 1998–2003. *Science* **282,** 682–689.
2. Botstein, D., White, D. L., Skolnick, M., and Davis, R. W. (1980) Construction of a genetic linkage map in man using restriction fragment length polymorphisms. *Am. J. Hum. Genet.* **32,** 314–331.

3. Weber, J. L. and May, P. E. (1989) Abundant class of human DNA polymorphisms which can be typed using the polymerase chain reaction. *Am. J. Hum. Genet.* **44,** 388–396.
4. Collins, F. S., Guyer, M. S., and Chakravarti, A. (1997) Variations on a theme: Cataloging human DNA sequence variation. *Science* **278,** 1580–1581.
5. Kruglyak, L. (1997) The use of a genetic map of biallelic markers in linkage studies. *Nat. Genet.* **17,** 21–24.
6. Nickerson, D. A., Taylor, S. L., Weiss, K. M., Clark, A. G., Hutchinson, R. G., Stengard, J., Salomaa, V., Vartiainen, E., Boerwinkle, E., Sing, C.F. (1998) DNA sequence diversity in a 9.7-kb region of the human lipoprotein lipase gene. *Nature Genet.* **19,** 233–240.
7. Sun, G. L., Diaz, O., Salomon, B., von Bothmer, R. (1999) Genetic diversity in Elymus caninus as revealed by isozyme, RAPD, and microsatellite markers. *Genome* **42,** 420–431.
8. Gusella, J. F., Wexler, N. S., Conneally, P. M., Naylor, S. L., Anderson, M. A., Tanzi, R. E., Watkins, P. C., Ottina, K., Wallace, M. R., and Sakaguchi, A. Y. (1983) A polymorphic DNA marker genetically linked to Huntigton's disease. *Nature* **306,** 234–238.
9. Risch, N. and Merikangas, K. (1996) The future of genetic studies of complex human diseases. *Science* **273,** 1516–1517.
10. Risch, N. and Teng, J. (1998) The relative power of family-based and case-control designs for linkage disequilibrium studies of complex human diseases. *Genome Res.* **8,** 1273–1288.
11. Larrey, D., Berson, A., Habersetzer, F., Tinel, M., Castot, A., Babany, G., Letteron, P., Freneaux, E., Loeper, J., and Dansette, P. (1989) Genetic predisposition to drug hepatotoxicity: role in hepatitis caused by amineptine, a tricyclic antidepressant. *Hepatology* **10,** 168–173.
12. Collins, F. S., Brooks, L. D., and Chakravarti, A. (1998) A DNA polymorphism discovery resource for research on human genetic variation. *Genome Res.* **8,** 1229–1231.
13. Christopoulos, T. K.(1999) Nucleic acid analysis. *Anal. Chem.* **71,** 425R–438R.
14. Hacia, J. G. (1999) Resequencing and mutational analysis using oligonucleotide microarrays. *Nat. Genetics Suppl.* **21,** 42–47.
15. Karas, M. and Hillenkamp, F. (1988) Laser desorption ionization of proteins with molecular masses exceeding 10,000 daltons. *Anal. Chem.* **60,** 2299–2301.
16. Wu, K. J., Steding, A., and Becker, C. H. (1993) Matrix-assisted laser desorption time-of-flight mass spectrometry of oligonucleotides using 3-Hydroxypicolinic acid as an ultraviolet-sensitive matrix. *Rapid Commun. Mass Spectrom.* **7,** 142–146.
17. Tang, K., Fu, D., Kötter, S., Cotter, R. J., Cantor, C. R., and Köster, H. (1995) Matrix-assisted laser desorption/ionization mass spectrometry of immobilized duplex DNA probes. *Nucleic Acids Res.* **23,** 3126–3131.
18. Jurinke, C., van den Boom, D., Jacob, A., Tang, K., Wörl, R., and Köster, H. (1996) Analysis of ligase chain reaction products via matrix assisted laser desorption/ionization time-of-flight mass spectrometry. *Anal. Biochem.* **237,** 174–181.

19. Jurinke, C., Zöllner, B., Feucht, H.-H., Jacob, A., Kirchhübel, J., Lüchow, A., van den Boom, D., Laufs, R., and Köster, H. (1996) Detection of Hepatitis B virus DNA in serum samples via nested PCR and MALDI-TOF mass spectrometry. *Genet. Anal.* **13,** 67–71.

20. Pieles, U., Zurcher, W., Schar, M., and Moser, H. E. (1993) Matrix-assisted laser desorption ionization time-of-flight mass spectrometry: a powerful tool for the mass and sequence analysis of natural and modified oligonucleotides. *Nucleic Acids Res.* **21,** 3191–3196.

21. Köster, H., Tang, K., Fu, D. J., Braun, A., van den Boom, D., Smith, C. L., Cotter, R. J., and Cantor, C. R. (1996) A strategy for rapid and efficient DNA sequencing by mass spectrometry. *Nat. Biotechnol.* **14,** 1123–1129.

22. Fu, D. J., Tang, K., Braun, A., Reuter, D., Darnhofer-Demar, B., Little, D. P., O'Donnell, M. J., Cantor, C. R., and Köster, H. (1998) sequencing exons 5 to 8 of the p53 gene by MALDI-TOF mass spectrometry. *Nature Biotechnol.* **16,** 381–384.

23. Berkenkamp, S., Kirpekar, F., and Hillenkamp, F. (1998) Infrared MALDI mass spectrometry of large nucleic acids. *Science* **281,** 260–262.

24. Braun, A., Little, D. P., and Köster, H. (1997) Detecting CFTR gene mutations by using primer oligo base extension and mass spectrometry. *Clin. Chem.* **43,** 1151–1158.

25. Braun, A., Little, D. P., Reuter, D., Muller-Mysock, B., and Köster, H. (1997) Improved analysis of microsatellites using mass spectrometry. *Genomics* **46,** 18–23.

7

Ink-Jet-Deposited Microspot Arrays of DNA and Other Bioactive Molecules

Patrick Cooley, Debra Hinson, Hans-Jochen Trost, Bogdan Antohe, and David Wallace

1. Introduction

The creation of microspot arrays of bioactive materials has been vigorously pursued by a number of research organizations and commercial companies in the past decade *(1,2)*. In addition to the high density and/or small area achieved by using microspot arrays, increased sensitivity is possible for some applications *(3)*. Ink-jet printing technology has been used to produce microspot arrays of both DNA and antibodies *(4)*, and the procedure to create oligonucleotide microspot arrays is described in this chapter.

1.1. Drop-On-Demand Array Ink-Jet Printhead

A 10-channel demand mode ink-jet printhead is utilized to deposit oligonucleotide probes onto the substrate. Each channel has its own fluid inlet, allowing the deposition of 10 bioactive solutions simultaneously. The printhead is a single integrated structure, eliminating the need to align the placement of each of the 10 channels relative to each other. The printhead was adapted from a design developed for a high-speed/high-quality office printer *(5,6)*. A brief description of the construction and operating principle of the printhead is given next.

A multilayer piezoelectric block is created to form the integrated structure of the printhead. Using a diamond saw, precision grooves are placed in the block. For the printhead employed herein, 20 grooves are placed 1.0 mm apart and the groove dimensions are 140 µm (width) by 400 µm (depth). For the office printer configuration, 122 grooves are sawn in the block on 170-µm centers, and groove dimensions are 85 µm (width) by 360 µm (depth). After the

From: *Methods in Molecular Biology, vol. 170: DNA Arrays: Methods and Protocols*
Edited by: J. B. Rampal © Humana Press Inc., Totowa, NJ

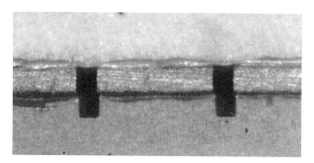

Fig. 1. End view displaying two channels of the 10-channel printhead before the orifice plate is attached.

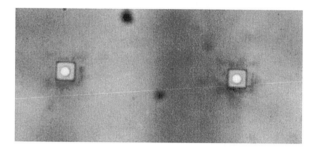

Fig. 2. Close-up view of orifice plate revealing two 40-μm-diameter orifices.

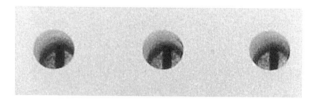

Fig. 3. A view of the cover plate displaying the holes created for the fluidics interconnection to the channels.

grooves are sawn, a cover is placed over the grooves to form enclosed rectangular channels for the working fluids. **Figure 1** presents an end view of two channels.

Next, a polymer orifice plate containing nominally 40-μm-diameter orifices is attached to the end of the channels as shown in **Fig. 2**. To complete the printhead, the electrical and fluidic interconnections are added. **Figure 3** shows the fluidic interface to the channels through the cover. Every other groove has

Fig. 4. Ten-fluid demand mode array printhead used to dispense oligonucleotide solutions.

a fluidic interface, resulting in 10 active channels on 2-mm centers. **Figure 4** illustrates a completely assembled printhead.

In forming the fluid channels by sawing groves in the multilayer piezoelectric block, the walls are also formed into actuators. These actuators are addressed through the electrical interconnections. When voltages are applied to two walls, they move as shown in **Fig. 5** owing to the piezoelectric properties of the block material. This motion results in nominally 40-μm-diameter droplets being formed, as shown in **Fig. 6**.

2. Materials

2.1. Printhead Setup

2.1.1. Fluid Preparation

1. Oligonucleotide solutions (*see* **Note 1**).
2. Ethylene glycol (99+%, spectrophotometric grade) (Aldrich, Milwaukee, WI).
3. Ultrafree MC 0.22 μm filter unit (Millipore, Bedford, MA).
4. Vacuum desiccator (Bel-Art).

2.1.2. Channel Priming and Fluid Loading

1. Syringe (10 μL) (Becton Dickinson, Franklin Lakes, NJ).
2. Cameo 25-mm syringe filters (nylon, 5 μm) (MSI, Westboro, MA).

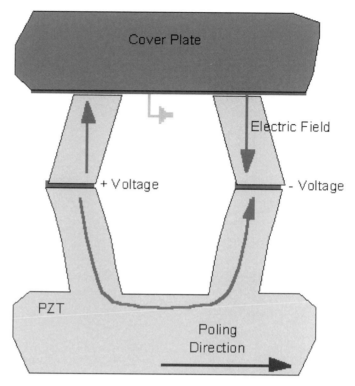

Fig. 5. Schematic drawing displaying a cross-section of a channel in an array printhead. The outward movement of the walls of the channel from vertical is the result of a shear mode displacement of the PZT when an electric current is applied. The resulting increase in the cross-sectional area of the channel creates an acoustic wave leading to the ejection of droplets.

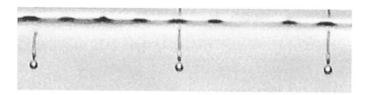

Fig. 6. Photograph of 40-μm drops being dispensed by every fourth channel from a 120-channel array printhead. Distance between drops is 510 μm.

3. Double-distilled water (ddH₂O).
4. Clean room swabs, foam material (Fisher, Pittsburgh, PA).
5. Pipettor (10 μL) (Rannin, Woburn, MA).
6. Pipet tips (10 μL) (VWR, West Chester, PA).

2.1.3. Jetting Setup

1. Humidifier (Duracraft, Whitinsville, MA).
2. Clean room swabs, foam material (Fisher).
3. Syringe (10 µL) (Becton Dickinson).
4. Cameo 25-mm syringe filters (nylon, 5 µm) (MSI).

2.1.4. Cleaning

1. Isopropyl alcohol (99% [v/v]) (VWR).
2. ddH_2O.
3. Eliminase (Fisher).
4. Vacuum source.

2.2. Printing System Setup

2.2.1. Array Printing Patterns

1. Excel software (Microsoft).

2.2.2. Spot Size Adjustment

1. Water-sensitive paper (TeeJet, Wheaton, IL).
2. Glass slides (Gold Seal, Portsmouth, NH).
3. Microscope.
4. Micrometer (Olympus, Tokyo, Japan).

2.2.3. Spot-To-Spot Alignment

1. Water-sensitive paper (TeeJet).
2. Glass slides (Gold Seal).
3. Microscope with charge-coupled device (CCD) camera (Kappa, Monrovia, CA).
4. Optimas image analysis software (Optimas, Bothell, WA).

2.2.4. Substrate Handling

1. Glass slides (Gold Seal).
2. Incubator.
3. Slide boxes (Ted Pella, Redding, CA).
4. Desiccant (N. T. Gates, Camden, NJ).

2.2.5. Substrate Loading

1. Vacuum chuck (per user requirements).
2. Vacuum pump, continuous operation (Gast, Benton Harbor, MI).
3. Custom glass slides, optional (Gold Seal).

2.2.6. Quality Control

1. Rhodamine dye (Molecular Probes, Eugene, OR).
2. Fluorescent microscope with CCD camera (Kappa).

2.3. Printing System

The basic concept of our print systems is to move the substrate underneath the printhead; when it is in position, the head is commanded to dispense one or more droplets. Two modes of observation are implemented: survey of the printed pattern using a downward-looking camera or microscope mounted close to the printhead, and observation of drops in flight by placing a strobed light-emitting diode in front of the printhead and a CCD camera behind it.

Our print system contains up to four axes of motion. Always present is an x-y motion system to carry and move the printing substrate underneath the printhead. A vacuum chuck having alignment pins is used to secure flat substrates, such as glass slides. A small vertical axis, labeled z, is used to raise and lower the printhead. The fourth axis (labeled w) is used to move the rear camera for observing drops in flight transverse to the direction of view.

The piezoelectric element of the printhead receives voltage pulses from a pulse generator to dispense droplets on demand. Rectangular pulses of tens of microseconds in length and tens of volts to about 100 V are commonly used. We found it beneficial to allow for an opposite second component to form a bipolar pulse, and to allow control of the rise and fall times. Thus, our pulse generators are simple versions of an arbitrary waveform generator, into which an analog output amplifier is integrated. For our multichannel printheads, we currently multiplex the output of a single pulse generator.

The print system is controlled and operated from a personal computer (PC). **Figure 7** is a schematic showing the various components and communication links in the print system. A single control program allows the user to set operating parameters and select and execute printing patterns. The operating parameters include those describing the voltage, frequency and shape of the waveform (rise/fall, dwell and echo pulse times), and the speed and acceleration of the motion stages. Some simple spot array printing patterns (lines, borders, arrays) are defined internally. Other spot array printing patterns can be provided in text data files, such as Excel. One form is a list of points with some additional information on a point-by-point basis, such as how many drops to dispense at the given location and which channel should be dispensing the drops.

An additional feature of our research printing system is the ability to survey printed patterns through a fluorescence microscope outfitted with a CCD camera linked to the PC. We are using the Optimas program version 6.2 for image analysis. The print station control program can use Optimas via DDE connection to automatically capture and analyze images. The control program then converts the results (locations and areas on screen) into measurements of actual positions and areas of the printed spots. The transformation between the screen

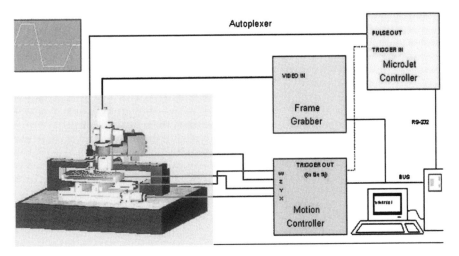

Fig. 7. Schematic illustrating the various components and communications links of the print system used to create oligonucleotide spot arrays.

and substrate scales is determined by observing the motion of a test spot on the screen for specific motions of the substrate.

3. Methods

The methods described herein are strictly a description for the deposition of oligonucleotide-containing solutions to create microarrays using ink-jet technology. The description for the synthesis of the oligonucleotides, glass slide preparation, hybridization conditions, and spot signal development is described in Chapters 9 and 11.

3.1. Ten-Fluid Printhead Setup

3.1.1. Fluid Preparation

Fluid properties have the greatest influence in terms of the jetability of biological solutions. Biological fluids vary in terms of wetting nature, bubble formation, viscosity, surface tension, rate of drying, and particulate content. These variations are owing to the buffer content and concentration, the concentration of biological material (DNA, protein, and enzymes), and the cleanliness of the fluid. Modifications of a biological solution may be necessary if jetting is problematic when attempting to dispense. This may require decreasing the concentration of the buffer or biological material in the solution. The use of a modification agent such as dimethyl sulfoxide or ethylene glycol *(7)* may also be required. For our system, we have found the addition of ethylene glycol to be most effective for improving jetability with the least effect on the integrity of the biological material (DNA) and postprocessing (oligonucleotide probe attachment and hybridization).

1. The oligonucleotide solutions should contain 20% (v/v) ethylene glycol (*see* **Note 2**). Bring the solutions to room temperature. Temperature changes will alter the fluidic properties of the solutions and affect the jetting behavior.
2. Filter the solutions using a nonsterile Ultrafree MC 0.22-µm microcentrifuge filter unit (*see* **Note 3**).
3. Degas the fluids for 15 min using a vacuum desiccator. Degassing of the fluids will reduce bubble formation in the printhead.

3.1.2. Channel Priming and Fluid Loading

1. Prime the channel to be loaded 3–5 s with ddH$_2$O using a syringe with a 5-µm filter attached (*see* **Note 4**).
2. Wipe the excess ddH$_2$O from the orifice plate using a swab.
3. Using a pipettor with a 10-µL pipet tip, load a volume of 10 µL of oligonucleotide solution into the manifold inlet until a small bead of fluid appears at the orifice. Continue to maintain the plunger in the depressed position.
4. While the pipettor plunger is depressed, carefully remove the pipet tip from the manifold inlet.
5. Carefully wipe the bead of fluid remaining at each orifice. If using a different oligonucleotide solution for each channel, use a new swab for each channel to avoid cross contamination. To avoid cross contamination of fluids from the adjacent channels, the wiping action should be done perpendicular to the long axis of the orifice plate (*see* **Note 5**).
6. Repeat this procedure for the remainder of the channels and fluids.

3.1.3. Jetting Setup

1. Utilize a humidifier to maintain a 40–50% relative humidity in the working area in which the printhead will be operated. This humidity requirement minimizes jetting failure by reducing fluid drying at the orifice.
2. Position the printhead to provide a perpendicular view of the orifice for drop observation using the video system.
3. Apply an electronic drive pulse with waveform parameters as specified in the printhead specification test form for the particular channel being operated. The waveform frequency for printing is usually 60 Hz.
4. Adjust the strobe delay to position drop formation outside of the orifice for observation.
5. If jetting does not occur, wipe the orifice using a swab. If jetting continues not to occur, adjust waveform parameters in an attempt to initiate jetting (*see* **Note 6**).
6. If jetting is evident, observe the drop formation for stability. Adjusting the volt-

age and dwell time can stabilize a drop that is bouncy or intermittent (*see* **Note 7**). Adjusting the voltage and dwell time can also eliminate satellite formation (*see* **Note 8**).

3.1.4. Cleaning

1. Aspirate oligonucleotide solutions from all channels of the printhead.
2. Lower the printhead into a well containing ddH$_2$O and briefly aspirate each of the channels.
3. Repeat **step 2** using isopropyl alcohol.
4. Immerse the printhead into the Eliminase and briefly aspirate each of the channels. Soak for 5 min.
5. Rinse the printhead well using ddH$_2$O.
6. Immerse the printhead orifice plate in ddH$_2$O and aspirate each of the channels until the fluid moving through the vacuum tubing is no longer foamy.
7. Immerse the printhead orifice plate in isopropyl alcohol and aspirate each of the channels.
8. Remove the printhead from the isopropyl alcohol and aspirate each of the channels until dry.

3.2. Printing System Setup

3.2.1. Array Printing Patterns

Pattern files for printing spot arrays using the printing system described in **Subheading 2.2.** are written using Excel. The instructions for writing a pattern file are briefly outlined next:

1. Enter in column 1 the values for the *x*-axis location for all array spots.
2. Enter in column 2 the values for the *y*-axis location for all array spots.
3. Enter in column 3 the fluid number, which corresponds to the channel number.
4. Enter in column 4 the number of drops to be dispensed per spot location.
5. Enter in column 5 a number 1 for a pause to occur at the desired location during the printing run. Enter a 0 if no pause is required.
6. Save the recipe as a comma delimited (csv) file.

3.2.2. Spot Size Adjustment

The wetting properties of the fluid and the wetting nature of the substrate control spot size. Larger spots are the result of a hydrophilic substrate and/or a wetting solution containing surfactant materials (buffers and proteins). Smaller spots are achieved when using a hydrophobic substrate (silinated glass), and a less wetting solution, such as one containing ethylene glycol.

1. Load a water-sensitive paper attached to a glass slide onto the substrate holder on the stage of the print system (*see* **Note 9**).
2. Adjust the substrate-to-orifice plate distance to approx 200 µm (*see* **Note 10**).

3. Print the desired spot array having three drops per spot.
4. Observe the spot size for each of the channels that have dispensed drops. Measurements can be made using a microscope with a micrometer.
5. Reduce or increase the number of drops per spot to equilibrate spot size for all the channels.
6. Repeat printing of the spot array and observe the spot size.
7. Repeat **steps 5** and **6** until the spot sizes are satisfactorily equilibrated for all the channels.

3.2.3. Spot-To-Spot Alignment

1. Print an array of spots on 0.5-mm pitch from channels 1 through 10 onto water-sensitive paper. This array will be used to calculate the x- and y-axis offset changes required to obtain uniform spot-to-spot placement.
2. Position the printed array under the microscope with a CCD camera and use the Optimus software to calculate spot-to-spot placement.
3. Calculate the x- and y-axis offset required to correct any misalignment of the array spots.
4. Adjust the x- and y-axis location offsets in the motion control software.
5. Reprint the array and confirm spot-to-spot alignment.
6. The array printing process is now ready to commence once the substrates have been loaded onto the substrate holder.

3.2.4. Substrate Handling

Glass slides are prepared as described by Beattie et al. *(8)*. Prepared glass slides are incubated at 70°C 24 h prior to use. The printed slides are placed into slide boxes and stored in plastic bags with desiccant until processed.

3.2.5. Substrate Loading

Although the speed of the motion system largely controls the rate of throughput, the number of substrates held on the print stage is a significant limiting factor. We can currently load 18 glass slides onto our printing stage. Throughput can be increased to 92 per print run using 15×25 mm custom glass slides. However, for ease of handling, most users opt for the standard size glass slide.

The glass substrates are affixed to the stage via a vacuum chuck. This chuck has a series of guide pins to aid in the position alignment of the slides. A series of holes in the base of the chuck allow the vacuum to hold the slides in place securely. To provide for increased throughput, this chuck can be interchanged with another that has been previously loaded with slides.

3.2.6. Quality Control

We have found it useful to incorporate a small quantity of rhodamine dye (*see* **Note 12**) in the oligonucleotide solutions during printing. As shown in

Fig. 8. Photograph of an array with spots containing rhodamine dye when illuminated by UV light with an excitation wavelength of 550 nm. Spot diameters are 65 μm printed on 200-μm centers.

Fig. 8, an inspection can be performed under ultraviolet (UV) illumination to determine whether all the spots have been deposited, their placement accuracy, and spot size (diameter) uniformity. This inspection is conducted on the print station using a survey system controlled by software to automatically collect confirmation of spot deposition, placement accuracy, and size. This software operates in conjunction with Optimas image software to acquire images for analysis.

4. Notes

1. We have dispensed a wide variety of oligonucleotide solutions (5–75-mer) suspended in ddH$_2$O, and buffer solutions, such as NaHCO$_3$, LiCl, and K$_2$CO$_3$. The increased viscosity and reduced surface tension of concentrated buffer solutions can lead to improved jetting. However, solutions containing concentrated salts can have a precipitate accumulate near the orifice resulting in a blockage that may impede jetting.
2. The addition of 20% ethylene glycol increases the viscosity and provides a humectant capacity to protein-containing solutions. The increased viscosity improves jetting performance by reducing satellites, increasing drop stability, and reducing wetting of the orifice. The humectant nature of ethylene glycol reduces the vapor pressure and drying of the aqueous biological solutions. Thus, orifice clogging owing to protein concentration and precipitation (filming over) is reduced.

3. Cleanliness is of the utmost importance when attempting to dispense solutions through a small orifice. A dust-free working environment provided by HEPA laminar flow is highly recommended. In addition, solutions used to flush the printhead channels (isopropyl alcohol/ddH$_2$O) should be delivered using a syringe (10 mL; Becton Dickinson) having a 5.0-μm filter attached (MSI).

4. Priming of the channels within the printhead using ddH$_2$O displaces the air in the channel, thus enabling aqueous-based solutions to wet with the surfaces of the channel. Air bubbles can be trapped if the wetting of the fluid with the surfaces of the channel is not uniform. These air bubbles will impede jetting by attenuating the acoustic wave generated by the lead zirconate titanate (PZT).

5. The cross-contamination of solutions is always a hazard when working with a printhead. Label the manifold inlets to correspond to solutions being loaded to avoid confusion. Be sure not to purge excess solution through the printhead channels. Excess solution can migrate across the orifice plate and contaminate fluid in adjacent channels. Be sure to change the tips on syringes and vacuum tubing when changing between channels during flushing and aspiration procedures.

6. Increasing the drive voltage in 10-V increments can be used to induce jetting. However, if jetting does not occur after an increase of 40–50 V, the channel should be reloaded with the oligonucleotide solution (air bubbles may be present). Aspirate the oligonucleotide solution from the channel and prime the channel using ddH$_2$O delivered by a syringe. Then, swab the excess ddH$_2$O from the orifice and reload the oligonucleotide.

7. Increasing the drive voltage by 3–7 V and/or changing the dwell time ± 3–5 μs will often stabilize drop formation. If drop formation stability cannot be achieved by this method, try repriming and reloading the channel.

8. Increasing the dwell time and/or reducing the voltage can eliminate a trailing satellite located above the main drop. Reducing the dwell time can eliminate a satellite located below or in orbit with the main drop.

9. The water-sensitive paper is available in 25 × 75 mm sheets and can be attached to a glass slide using double-sided tape. The paper is yellow and turns a dark blue when the surface is in contact with water. To prevent contact of the paper with the face of the printhead, the paper should be attached to the glass slide, as flat as possible. Gloves should be worn when handling the paper to avoid blemishing the surface.

10. For best results the printhead should be as close as possible to the substrate during printing. This distance is governed by the flatness of the substrate.

11. Spot-to-spot misalignment in an array is more obvious as the center-to-center distance of the spots is reduced. A spot-to-spot misalignment of up to ±25 μm can be acceptable for arrays having spots on 500-μm pitch. Arrays having spots printed on 200-μm pitch or less may not tolerate a spot-to-spot misalignment of more than ± 10 μm.

12. We have found the addition of 0.5 m*M* rhodamine red maleimide provides a signal that can be observed using our fluorescent system, while having the least effect on oligonucleotide attachment and hybridization.

References

1. Wallace, D. B., Shah, V., Hayes, D. J., and Grove, M. E. (1996) Photo-realistic ink jet printing through dynamic spot size control. *J. Imaging Sci. Technol.* **40,** 390–395.
2. Pies, J. R., Wallace, D. B., and Hayes, D. J. (1993) High density ink jet printhead. US Patent 5,235,352.
3. Eichenlaub, U., Berger, H., Finckh, P., Karl, J., Hornauer, H., Ehrlich-Weinreich, G., Weindel, K., Lenz, H., and Sluka, P. (1998) *Microspot®—A Highly Integrated Ligand Binding Assay Technology,* Proceedings of the Second International Conference on Microreaction Technology, New Orleans.
4. Lipshutz, R. J., Fodor, S. P. A., Gingeras, T. R., and Lockhart, D. J. (1999) High density synthetic oligonucleotide arrays. *Nat. Genet. Microarray Suppl.* **21,** 20–24.
5. Ekins, R. (1994) Immunoassay: Recent developments and future directions. *Nucl. Med. Biol.* **21,** 495–521.
6. Wallace, D. (1996) Ink-jet based fluid microdispensing in biochemical applications. *Lab. Automation News* **1,** 6–9.
7. Banczak, D. P. and Tan, W. E. (1997) Jet printing ink composition. US Patent 4,021,252.
8. Beattie, W. G., Meng, L., Turner, S. L., Varma, R. S., Dao, D. D., and Beattie, K. L. (1995) Hybridization of DNA targets to glass-tethered oligonucleotide probes. *Mol. Biotechnol.* **4,** 213–225.

8

Printing DNA Microarrays Using the Biomek® 2000 Laboratory Automation Workstation

David W. Galbraith, Jiří Macas, Elizabeth A. Pierson, Wenying Xu, and Marcela Nouzová

1. Introduction

DNA microarray technologies have been developed as a high-throughput means to study transcriptional regulation (for a recent review, *see* **ref. *1***). Large numbers of DNA samples (either complete cDNAs or ESTs) are immobilized as very high-density arrays on solid surfaces, typically glass. Microarray analysis is conceptually similar to that of a reverse dot blot, whereby the tethered "probe" DNA samples comprising the array are hybridized with fluorescently labeled "targets" produced from mRNAs isolated from tissues of interest (for a discussion of terminology, *see* **ref. *2***). The microarrays are then scanned to quantitate the signals produced by the hybridization of the labeled targets to the individual array elements. In this way, parallel monitoring can be done for all transcripts represented in the array. Direct comparison of gene expression patterns between two tissues, or two different experimental conditions, is done by simultaneous hybridization of corresponding targets labeled with different fluorochromes. This approach has been successfully employed to study gene expression in both prokaryotes and eukaryotes (*1,3–10*).

The production of DNA microarrays can be done using instruments that are specifically designed for this purpose and that recently have been commercialized (*11*). Because these instruments were not readily available at the time this work was started, we employed a different strategy: that of adapting a generic robotic workstation for array fabrication.

The Biomek® 2000 Laboratory Automation is designed to perform repetitive laboratory operations, primarily liquid handling. The Biomek has very good spatial positioning capabilities and is user programmable. We employed

From: *Methods in Molecular Biology, vol. 170: DNA Arrays: Methods and Protocols*
Edited by: J. B. Rampal © Humana Press Inc., Totowa, NJ

these properties in modifying the workstation to deliver DNA samples at high densities onto glass microscope slides in the form of microarrays. In this chapter, we describe procedures for converting the Biomek for printing arrays, comprising as many as 3000 DNA elements.

2. Materials

2.1. Workstation Configuration

Our Biomek® 2000 Workstation (Beckman, Fullerton, CA) is equipped with right- and left-side modules (part no. 609047 and 609048), the 96 HDR System (Beckman, part no. 267616), and four Labware Holders (part no. 609120). It is installed in a laminar flow hood (Nuaire™ Class II type A/B3; Nuaire, Plymouth, MN). A deep well microtiter plate (Beckman, part no. 267006) is used to hold the microwell plate containing the DNA samples. The high-density replicating tool, tool rack, two reservoir modules, and fan unit come with the HDR system.

2.2. Printhead

The printhead is constructed from stainless steel and attaches to the base of the Biomek HDR tool. In our case, construction was done by Geometric, a subsidiary of Hi-Tech Machining & Engineering, L.L.C. (Tucson, AZ).

2.3. Software

Array printing by the Biomek workstation is controlled by a program written in Tool Command Language (Tcl) *(12)*, and is launched from the workstation control environment (Bioworks) operating under Microsoft Windows. The program script is available from us on request (also *see* http://latin.arizona.edu/galbraith/robot/).

2.4. Molecular Supplies

1. Polymerase chain reaction (PCR) amplification was done in an MJ Research Thermal Cycler (PTC-200 DNA Engine and 96V Alpha unit) using Concord microwell plates (MJ Research).
2. 5' C-6 amino-modified primers (Life Technologies, Rockville, MD).
3. dNTPs (Life Technologies).
4. AmpliTaq® Gold DNA polymerase (PE Applied Biosystems, Foster City, CA).
5. For hybridization, mRNA was isolated using the MicroPoly(A) Pure kit (Ambion, Austin, TX).
6. First-strand cDNA was synthesized using the SuperScript Preamplification System (Life Technologies).
7. For preparation of labeled targets, Cy3-dUTP and Cy5-dUTP were used (cat. no. PA 55022; Amersham, Pittsburgh, PA).

8. M13 reverse and T7 forward primers for insert amplification from pBluescript™ plasmids (Stratagene, La Jolla, CA).
9. T7 and SP6 primers for insert amplification of PZL1 plasmids (Life Technologies).
10. 2X Saline sodium citrate (SSC) buffer: 3 *M* NaCl and 30 m*M* Na citrate, pH 8.5.
11. PicoGreen (Molecular Probes, Eugene, OR).
12. Alcoholic sodium borohydride solution: 1.0 g of NaBH$_4$ dissolved in 300 mL of phosphate-buffered saline (PBS), to which 100 mL of 95% (v/v) ethanol is added.
13. Falcon disposable conical centrifuge tube (50 mL) (Falcon 352070; Becton Dickinson, Franklin Lakes, NJ).

2.5. Miscellaneous Supplies

1. Silylated slides for microarray fabrication were purchased from CEL Associates (Houston, TX). The slides are prescanned to check that the background fluorescence is acceptably low.
2. Sheets of Scott® absorbent wipes (Scott Paper, Philadelphia, PA).
3. Scotch™ Double Coated Tape (3M, St. Paul, MN).
4. STORM™ Phosphorimager/scanning fluorometer (Molecular Dynamics, Sunnyvale, CA).
5. ScanArray 3000 Microarray Scanner (GSI Lumonics, Billerica, MA).

3. Methods
3.1. Configuring the Workstation

1. Position the microwell plates and reservoirs on the Biomek® 2000 work surface using the four Labware holders (**Fig. 1**). Locate the individual components at the following positions (as defined for the Biomek): Position A1, fan; Position A2, reservoir containing 96% (v/v) ethanol; Position A3, deep-well microtiter plate (carries the microwell plate containing the DNA samples); Position B1, reservoir containing deionized water; Position B2, Concord PCR microwell plate covered with three sheets of Scott absorbent wipes. Place the PCR microwell plate containing the DNA samples on the work surface within the deep-well microtiter plate.
2. Align a glass plate (310 × 260 × 5 mm) with the right side of the worksurface (placed within the front and rear boundaries of the work surface), anchoring it using Scotch Double Coated Tape. Attach up to 28 silylated microscope slides to the upper surface of the glass plate using double-sided tape. The slides are arranged in four columns of seven slides, and are positioned with labels to the right side. The slide printing order is from back to front, starting in the left-hand column and then moving right.

3.2. Constructing and Configuring the Printhead (see Note 1)

1. Machine the printing pins from 9.5-mm- (3/8-in.) diameter 304 stainless steel rod. Cut the rod into 35-mm (1 3/8-in.) sections. Turn the rod to a diameter of 3.15 mm (0.124 in.) to a distance of 19 mm (3/4 in.). Machine this end to a conical tip subtending a 10° angle.

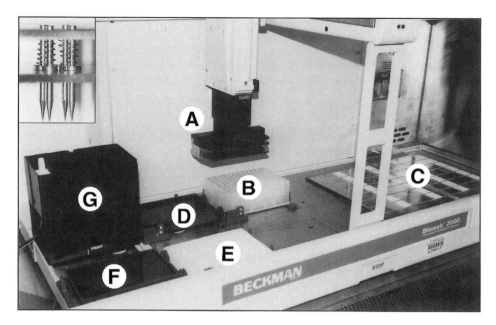

Fig. 1. The Biomek 2000 workstation equipped for printing DNA microarrays. Layout of individual components on the work surface. The slides are arranged in four columns on the right side of the worksurface. (**A**) Printing tool. (**B**) Microwell plate. (**C**) Slides. (**D**) Ethanol wash. (**E**) Paper pad. (**F**) Water wash. (**G**) Fan. (**Inset**) Detail of modified printing tool and pins.

2. Turn the opposite end of the rod to a diameter of 3.15 mm (0.124 in.), to a distance of 14.3 mm (9/16 in.). This leaves a 9.5-mm- (3/8-in.) diameter collar having a width of 1.7 mm (1/16 in.).
3. Cut a 0.125-mm (5/1000-in.) slit in the conical end of the rod to a depth corresponding to base of the cone, using Electrical Discharge Machining techniques.
4. Construct a two-plate assembly to carry the pins. Fabricate these two plates from 3.2-mm- (1/8-in.) thick stainless steel, having the same *x* and *y* dimensions as the Biomek HDR tool.
5. Space these apart using two 25.4 × 12.7 × 6.4 mm (1 × 1/2 × 1/4 in.) spacers. Hold these in place using the screws of the Biomek HDR tool.
6. Drill holes (3.2 mm diameter) in the center of the plates to form a 9 × 9 mm array. We have employed four holes, but larger numbers (8, 16, or 32) are theoretically possible (*see* **Note 2**). Prior to use, clean the pins by sonication in 0.1% sodium dodecyl sulfate (SDS) for 1 h, followed by washing in water, and in 96% ethanol.
7. Assemble the pin holder and printing pins, using springs taken from retractable ballpoint pens to seat the printing tips against the lower of the two plates.

3.3. Downloading and Running the Tcl Scripts (see Note 3)

1. Obtain the Tcl scripts from us (galbraith@arizona.edu). The scripts are copyrighted, but are freely available for noncommercial applications.
2. Copy the scripts to the Biomek workstation computer, and then load Bioscript.
3. Launch the scripts from the Bioworks environment according to the manufacturer's instructions.

3.4. Determination of Spotting Accuracy

To determine the accuracy of printing, arrays (12×8 elements) comprising 96 DNA samples dissolved in 2X SSC are spotted onto cover slips. Once dry, microarrays are projected at a 20-fold magnification onto a computer-designed print of an "'ideal" grid, and the distance of the grid marks from the actual centers of the sample dots is determined. The precision of printing is expressed as the standard deviation of positioning along the *x*- and *y*- axes.

3.5. Preparing DNA Samples for Spotting

1. Employ standard PCR techniques to amplify DNA samples using 5'-amino-modified primers. In our case, M13 reverse and T7 forward primers were used for insert amplification from pBluescript plasmids and T7 and SP6 primers were employed for insert amplification of PZL1 plasmids.
2. Following amplification, estimate the concentration of DNA in an aliquot ($1 \mu L$) of the product using PicoGreen according to the manufacturer's guidelines. Quantitate fluorescence using a STORM phosphorimager/scanning fluorometer.
3. Analyze PCR products ($1 \mu L$) using gel electrophoresis (1.5% [w/v] agarose in Tris-acetate-EDTA buffer, pH 8.5) to determine product size and confirm specificity of amplification.
4. Precipitate the PCR products, following the addition of 0.1 vol of 3 *M* sodium acetate (pH 5.2) and 1 vol of isopropanol, by centrifugation for 50 min at 2700*g*. Wash the pellet with 2 vol of 70% ethanol. Recover the pellet by centrifugation, and dry it.
5. Resuspend the PCR products in 2X SSC buffer to a final concentration of 0.2–0.3 µg/µL. We typically obtain 2–4 µg of PCR product from a 50-µL PCR reaction, providing a final resuspended volume of 10–15 µL. For optimal printing, at least 5–10 µL of DNA, with a final concentration of >0.2 µg/µL, is recommended. This is sufficient to produce at least 500 slides (*see* **Note 4**).

3.6. Printing and DNA Immobilization

PCR products, amino modified at their 5' ends, are printed on silylated microscope slides. After printing, samples must be immobilized onto the glass surface. Immobilization involves reaction of available primary and secondary amines with aldehydes present on the surface of the silylated slides. The resultant Schiff's bases are reduced, irreversibly linking the DNA to the glass surface. The following steps, previously described in **refs.** *9* and *10*, are appropriate only for immobilization of DNA using this type of surface chemistry (*see* **Note 5**).

1. Allow the printed slides to dry overnight (or for up to 4 d) in a slide box maintained in darkness at room temperature.
2. Place the slides in a humid chamber for 4 h at room temperature to rehydrate the arrays.
3. Wash the slides for 1 min in 40 mL of 0.2% SDS by repeatedly raising and lowering them in the solution using forceps. Follow this with rinsing (1 min) in 40 mL of distilled water.
4. Incubate the slides for 5 min in alcoholic sodium borohydride solution (1.0 g of $NaBH_4$ dissolved in 300 mL of PBS, to which is added 100 mL of 95% [v/v] ethanol).
5. Rinse the slides four times for 2 min each in 40 mL of distilled water. The slides can be stored dry and in darkness at 4°C for up to 12 mo (*see* **Note 6**).

3.7. Preparation of Labeled Targets and Array Hybridization (see Notes 7 and 8)

1. Isolate mRNA using the MicroPoly(A) Pure kit, or a similar product providing yields of 3 to 4 µg polyA⁺ RNA/g of fresh weight of tissue.
2. Synthesize first-strand cDNA using the SuperScript Preamplification System. We employ 2 µg of RNA sample, 5 µg of oligo dT primer, and 13 U/µL of Superscript II in a 30-µL reaction volume, and add dNTPs at the following final concentrations: dTTP (200 µ*M*), Cy3-dUTP or Cy5-dUTP (100 µ*M*), and dATP, dCTP, and dGTP (500 µ*M* each). Labeled targets can be stored in microfuge tubes in darkness at –20°C for up to 6 mo.
3. Define the target signal strength through direct printing, as described previously (the target will require dilution by approx 50-fold), followed by scanning, as described below.
4. Denature the immobilized DNA probes by immersing the slides for 2 min in distilled water at 92–95°C, then transferring them quickly to 96% ethanol for 20 s, followed by air-drying. This can be conveniently done by centrifuging the slides vertically oriented within a 50-mL Falcon disposable conical centrifuge tube for 1 min at 100 g.
5. Denature the fluorescently labeled targets by transferring the microfuge tubes to 100°C for 2.5 min, followed by immediate transfer onto ice.
6. Apply hybridization mix (5X SSC, 0.1% SDS) containing denatured target (8 µL) onto slides preheated to 65°C (*see* **Note 9**). Cover with a standard glass cover slip, and leave in a humid chamber at 65°C overnight. For the humid chamber, we employ a sealed glass desiccator placed in a standard incubator, humidifying by addition of 40 mL of 2X SSC to a small reservoir within the desiccator.
7. Wash the slides sequentially in 40 mL of 5X SSC and 0.1% SDS for 5 min, 40 mL of 0.2X SSC and 0.1% SDS for 5 min, and 40 mL of 0.2X SSC for 2 min at room

temperature. Remove excess liquid by centrifuging (2 min at 100g) the slides in 50-mL Falcon centrifuge tubes.
8. Immediately transfer the slides for scanning, using the ScanArray 3000.

4. Notes

1. Adapting the Biomek 2000 for printing DNA microarrays requires only minor modifications to the hardware and use of the appropriate software. The modifications (**Fig. 1**) to the workstation tool originally designed for bacterial colony replication are based on the design described by Brown (*[13]; see also* http:// cmgm.stanford.edu/pbrown/arrayer.html). The modifications to the tool body include precisely positioning the printing pins between two machined plates and seating these pins using small springs (**Fig. 1**). The mechanism of spotting *(13)* involves capillary adsorption of approx 2 μL of the DNA sample within the slotted tip of the individual pins. Small droplets (approx 5 nL) are transferred onto the glass slides when their surfaces are touched by the pin. The Biomek modifications allow sequential printing of up to 28 slides; further slides could be accommodated, but this would require alterations to the Tcl programs.
2. In this case, new Tcl programs would be required.
3. The software program to control the movements of the printing tool is written in the Tcl language. The program is designed to allow reliable and accurate spotting; in particular, this involves a two-step instruction for each movement of the robotic arm in the x-y plane. The first movement positions the arm at a defined point slightly displaced from the desired position of spotting. The second movement moves the arm to the precise position of spotting. This two-step process minimizes positional inaccuracies introduced by mechanical hysteresis associated with single large-scale movements of the robot arm. The program also includes washing steps to eliminate sample cross contamination at the printing pin. We have previously confirmed the effectiveness of the pin washing steps employed during the printing of the microarray, and the even signal intensity of hybridization to replicate array elements indicates that differences in the DNA amounts among replicate dots are minimal *(14)*.
4. Products can be stored in sealed microtiter plates at 4°C for up to 6 mo.
5. The spotting procedure starts with loading the sample from one well of the 96-well microwell plate (we support standard PCR microwell plates, within a polypropylene deep-well microwell plate). The sample is then spotted to a defined position on up to 28 slides. The remaining sample is then removed from the slot by touching the printing tip against a pad of dry paper. This is followed by sequential washing in water and ethanol. Most of the ethanol is removed by again moving the tip to the paper pad; residual ethanol is evaporated by moving the spotting tool to the fan station for 10 s. The tip then proceeds to load and spot the next sample. The positional accuracy of printing by the Biomek was as follows: for nominal x and y repeat values of 500 μm, the SD ($n = 768$) was ±52.6 μm (x-dimension) and ±63.7 μm (y-dimension). The limiting spotting density that can be achieved depends on array element diameters. The printing pins generally

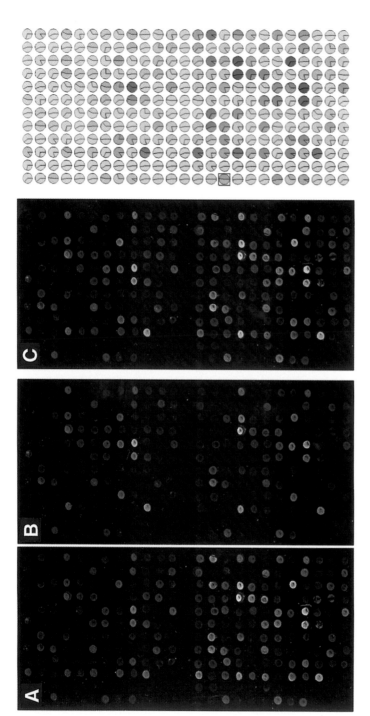

Fig. 2. Analysis of DNA microarrays of Arabidopsis *P450* genes. Array elements, produced from PCR products for a large cadre of Arabidopsis *P450* genes and various controls, are spaced at 500-μm intervals. Genes were identified in the form of cDNAs and ESTs in the available databases and were obtained from the Arabidopsis stock center. Hybridization was done using targets isolated from leaf and root tissues, labeled respectively with Cy3 and Cy5. Scanning of the microarrays at the two different wavelengths provides two TIFF images that can be merged to provide a pseudo-colored image, in which the relative proportion of expression of the genes under the two conditions is reflected by the resultant color (red, green, or yellow) of the array element. This information is also exported in quantitative form and can be used to create other visual representations such as pie diagrams. The majority of the genes are preferentially expressed in leaf tissues, possibly reflecting the predominant source tissue of the ESTs and cDNAs. (A) Cy3-channel scanned image (16-bit TIFF file, with 20-μm pixels). (B) Cy5-channel scanned image. (C) Pseudo-color image (8-bit RGB TIFF file, merged using Adobe Photoshop). (D) Pie diagram produced by ImaGene software (BioDiscovery, Los Angeles, CA).

gave spot sizes of about 125 µm. We were able to achieve reliable arraying with array element spacings ≥500 µm (**Fig. 2**). This is not as great a density as can be achieved with custom arraying instruments (an element spacing of ~200 µm *[1,15]*), but nevertheless allows immobilization of up to 3000 DNAs on a single microscope slide, which is sufficient for many research and diagnostic purposes.

6. In terms of rates of spotting, a single cycle (washing, drying, sample uptake, and array element printing on 28 slides) takes about 90 s. This compares to a cycle time for the GeneMachines Omnigrid Arrayer, spotting 100 slides, of about 56 s.

7. The microarrays prepared by the Biomek 2000 have been tested via hybridization of various controls *(14)*. **Figure 2** illustrates hybridization of arrays comprising Arabidopsis cytochrome P450 genes and various control genes using a mixture of Cy3- and Cy5-labeled targets derived from mRNA extracted from leaf and root tissues. We conclude that the arrays produced by the Biomek 2000 can be employed to provide useful biological information.

8. In terms of manufacturing, the printing pins and all parts needed for constructing the printing tool can be locally manufactured in a machine shop, at a cost of about $1000. Because the Biomek 2000 is designed for a large variety of liquid transfers, our modifications imply automation should be possible for the whole process of making DNA microarrays (colony picking, plasmid/PCR preparation, DNA arraying) within a single instrument. Further automation of the processes of hybridization, scanning, and analysis can be envisaged. Detailed instructions concerning alignment of the instrument prior to printing are available at our web page (http://latin.arizona.edu/galbraith/robot/).

9. It is important that slides be maintained at 65°C throughout the hybridization process.

Acknowledgments

This work was supported by grants from the U.S. Department of Agriculture (96-35300-3777) and the U.S. Army Research Office (DAAG559710102). It was also supported by NSF grants 9813360 (H. Bohnert, P.I.) and 9872657 (V. Walbot, P.I.) under the Plant Genome Initiative.

References

1. Duggan, D. J., Bittner, M., Chen, Y., Meltzer, P., and Trent, J. M. (1999) Expression profiling using cDNA microarrays. *Nat. Genet.* **21,** 10–14.
2. Phimister, B. (1999) Going global. *Nat. Genet.* **21,** 1.
3. Castellino, A. M. (1997) When the chips are down. *Genome Res.* **7,** 943–946.
4. Chu, S., DeRisi, J., Eisen, M., Mulholland, J., Botstein, D., Brown, P. O., and Herskowitz, I. (1998) The transcriptional program of sporulation in budding yeast. *Science* **282,** 699–705.
5. DeRisi J., Penland, L., Brown, P. O., Bittner, M. L., Meltzer, P. S., Ray, M., Chen, Y., Su, Y. A., and Trent, J. M. (1996) Use of a cDNA microarray to analyse gene expression patterns in human cancer. *Nat. Genet.* **14,** 457–460.

6. DeRisi, J. L., Iyer, V. R., and Brown, P. O. (1997) Exploring the metabolic and genetic control of gene expression on a genomic scale. *Science* **278,** 680–686.

7. Heller, R. A., Schena, M., Chai, A., Shalon, D., Bedilion, T., Gilmore, J., Woolley, D. E., and Davis, R. W. (1997) Discovery and analysis of inflammatory disease-related genes using cDNA microarrays. *Proc. Natl. Acad. Sci. USA* **94,** 2150–2155.

8. Schena, M. (1996) Genome analysis with gene expression microarrays. *BioEssays* **18,** 427–431.

9. Schena M., Shalon, D., Davis, R. W., and Brown, P. O. (1995) Quantitative monitoring of gene expression patterns with a complementary DNA microarray. *Science* **270,** 467–470.

10. Schena, M., Shalon, D. Heller, R., Chai, A., Brown, P. O., and Davis, R. W. (1996) Parallel human genome analysis: Microarray-based expression monitoring of 1000 genes. *Proc. Natl. Acad. Sci. USA* **93,** 10,614–10,619.

11. Bowtell, D. D. L. (1999) Options available—from start to finish—for obtaining expression data by microarray. *Nat. Genet.* **21,** 25–32.

12. Macas, J., Nouzova, M., and Galbraith, D. W. (1998) Adapting the Biomek 2000 Laboratory Automation Workstation for printing DNA microarrays. *BioTechniques* **25,** 106–110.

13. Shalon, D., Smith, S. J. and Brown, P. O. (1996) A DNA microarray system for analyzing complex DNA samples using two-color fluorescent probe hybridization. *Genome Res.* **6,** 639–645.

14. Ousterhout, J. K. (1994) Tcl and the Tk Toolkit, Addison-Wesley, Reading, MA.

15. Chung, V. G., Morley, M., Aguilar, F., Massimi A., Kucherlapa R., and Childs, G. (1999) Making and reading microarrays. *Nat. Genet.* **21,** 15–19.

9

Hybridization Analysis of Labeled RNA by Oligonucleotide Arrays

Ulrich Certa, Antoine de Saizieu, and Jan Mous

1. Introduction

Miniaturization and high-throughput parallel analysis are new concepts entering modern molecular biology. An exciting example of such technology originally developed by physicists and applied to biology is the microchip. The semiconductor industry manufactures silica chips with increasing numbers of smaller and smaller features, which allows incredible numbers of operations within split seconds. The same principle of performing multiparallel operations on a miniaturized solid phase has led to the development of miniaturized arrays of several hundreds of thousands of DNA fragments on a small chip. Arraying of DNA or RNA samples onto nitrocellulose or nylon membranes is a very common analytical procedure in a molecular biology laboratory. DNA chips or microarrays are miniaturized versions of these classical filter-hybridization techniques. With great precision, thousands of DNA fragments or DNA oligonucleotides can be printed onto glass chips or microscope slides. In the case of oligonucleotides, they can also be synthesized directly on coated silica chips.

The fabrication of high-density arrays on chips of approx 1 cm^2 for the parallel analysis of several thousand genes can be achieved by different techniques. cDNA clones or polymerase chian reaction fragments can be deposited on a solid support by microprinting or piezojet dispensing techniques *(1,2)* (*see* **Note 1**). The most elegant way to attach DNA fragments covalently onto the glass chips has been developed by Fodor et al. *(3,4)* and Pease et al. *(5)* Affymetrix (Santa Clara, CA). This method allows light-directed synthesis of several hundreds of thousands of different oligonucleotides in precise locations on the microchip. First, linkers modified with a photochemically removable protecting group are attached to the solid substrate. Light is directed

From: *Methods in Molecular Biology, vol. 170: DNA Arrays: Methods and Protocols*
Edited by: J. B. Rampal © Humana Press Inc., Totowa, NJ

through a photolithographic mask, illuminating specific grid squares on the chip to induce photodeprotection, or the removal of the blocking group, in those squares (features). The chip is then incubated with a nucleotide harboring a photolabile protecting group at the 5' end. The cycle continues: the chip is exposed to light through the next mask, which activates new grid squares for reaction with the subsequent modified nucleotide. Maximally four masks are needed to synthesize the four possible nucleotides at each level of the nascent oligonucleotide. Using the proper set of masks and chemical steps, it is possible to construct a defined collection of oligonucleotides, generally 20–25 bases long, each in a predefined position on the array. Currently, a standard chip can be packaged with about 400,000 individual, well-defined oligonucleotides representing many thousand genes.

The DNA chip technology is currently being applied in the areas of monitoring gene expression, polymorphism analysis, gene mutation analysis, and DNA sequencing.

The levels and the timing of gene expression determine the fate of the cells, their reproduction, differentiation, function, communication, and physiology. Measuring mRNA levels from all genes expressed in cells will help increase our understanding of these complex molecular processes. High-density arrays of oligonucleotide probes represent a prototype method for mRNA expression monitoring. Sequence information about the expressed genes or all genes of a particular genome is used directly to select oligonucleotides and to design the photolithographic masks for combinatorial synthesis of the probes on the derivatized glass as already described (*see* also **Fig. 1**). The arrays are intentionally redundant, because they contain collections of pairs of probes for each of the RNAs being monitored. Each probe pair consists of a 20- to 25-mer oligonucleotide that is perfectly complementary to a subsequence of a particular mRNA and a mismatch oligo that is identical except for a single base difference in a central position, which serves as an internal control for hybridization efficiency. The array hybridization experiments described in this chapter result in light intensity patterns monitored by a confocal laser scanner that can be interpreted in terms of gene identification and the exact and relative amounts of each transcript in a given cell or tissue. This technology has been successfully applied for monitoring gene expression in mammalian cells (*6*), yeast (*7*), and bacteria (*8*), measuring expression levels of less than one copy of mRNA to several hundred copies per cell in one hybridization experiment.

Evidently, studying biochemical pathways in such molecular detail is a fascinating application of this technology. On the other hand, gene chips can also be used to study and evaluate the pharmacological activity as well as toxic side effects of therapeutic drugs, both in vitro and in vivo. Different compounds may induce different biochemical reactions in target cells or tissues. There-

1. Oligonucleotide Sequence Selection

gtggacagatgacccgatagatgatgagagagagagagagatttttttgagatcacgatagtgacggatttaccgat

↓   ↓   ↓

gtggacagatgaccc   gagagagagagagattt   tagtgacggatttaccgat

2. Photochemical Oligonucleotide Synthesis

CHIP

3. Probing the Chip

labeled cRNA   labeled cRNA

4. Analysis

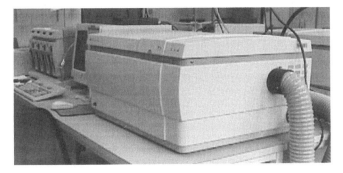

Fig. 1. Basic steps of gene expression monitoring using high-density oligonucle-otide arrays. 1, Selecting oligonucleotide sequences to represent the different genes of interest; 2, light-directed synthesis of oligonucleotides on the derivatized glass; 3, pre-paring fluorescent labeled cRNA from cells or tissues and hybridizing to the oligo-nucleotides on the chip; and 4, measuring the amount of cRNA specifically bound to oligonucleotide probes by confocal laser scanning technology.

fore, a fingerprint of clearly defined changes in gene transcription can be used to define a specific pharmacological activity or to indicate possible unwanted side effects. These molecular profiles not only lead to a better understanding of the biochemistry of a specific drug activity, but also can be used as surrogate markers during drug development.

High-density oligonucleotide probe arrays can also be applied to a broad range of nucleic acid sequence analysis problems, including pathogen identification, single-nucleotide polymorphism (SNP) detection, gene mutation analysis, and sequence checking. However, these applications are not discussed in detail herein.

In this chapter, we discuss and present in detail the materials, methods, and basic protocols for gene expression monitoring studies using oligonucleotide probe arrays.

2. Materials
2.1. GeneChip® Probe Arrays

The oligonucleotide arrays are manufactured and distributed by Affymetrix. Chips are shipped on wet ice and should be stored at 4°C. Chips cannot be frozen or reused because leakage of the hybridization chamber is likely to occur. The current shelf life of Affymetrix chips is 6 mo.

2.1.1. Eukaryotic Chips

Currently various formats of human, mouse, rat, and yeast chips are commercially available (*see* **Table 1**). The list of commercially available chips is growing continuously and chips with nematode, *Drosophila*, and *Escherichia coli* genes are planned for release in 2000. In addition, chips customized for certain applications such as cancer research or toxicology are being produced. Apart from commercial designs, customers have the option to design their own chips (*see* **Subheading 2.1.2.**). For the development of eukaryotic technology, we have designed a chip with a set of 1200 human, mouse, and rat genes.

2.1.2. Prokaryotic Chips

A custom designed microarray has been used to develop protocols for monitoring gene expression in prokaryotes: a low-density Affymetrix chip containing 100 *Streptococcus pneumoniae* genes and 100 *Haemophilus influenzae* genes (*8*). The second generation is a high-density chip containing 4880 genes representing the complete genomes of both *H. influenzae* and *S. pneumoniae*. Twenty-five probe pairs represent each gene unless the probe selection procedure did not identify 25 suitable oligonucleotides. The probe sequence is antisense for direct hybridization with labeled RNA.

Table 1
Commercial Oligonucleotide Arrays[a]

Organism	Number of genes represented	Description	Chip type[b]
Human	~35,000	Hu35K Set A, B, C, D no. 900184-87	High density
Human	~6000	Hu Gene FL Array no. 900183	High density
Human	~6000	Hu Gene FL Set A, B, C, D no. 900180	Low density
Mouse	~19,000	Mu19K Set A, B, C no. 900190-92	High density
Mouse	~11,000	Mu11K Set A, B no. 900188-89	High density
Mouse	~6500	Mu6500 Set A, B, C, D no. 900161	Low density
Rat	~19,000	U34 Set A, B, C no. 900249	High density
Rat	~850	Rat Toxicology U34 Array no. 900252	High density
Yeast	~6100[c]	Ye6100 Set A, B, C, D no. 900162	Low density

[a]For more information, *see* www.affymetrix.com.
[b]High-density utilize 24×24 μ features and low-density chips 50×50 μ features.
[c]Complete genome chip.

2.2. Isolation of RNA

2.2.1. Isolation of Bacterial RNA

1. Acid phenol (Gibco-BRL).
2. Proteinase K (Ambion).
3. RNase-free DNase (Promega).
4. RQ1 buffer (Promega).
5. NAE buffer: 50 mM sodium acetate, pH 5.1, 10 mM EDTA.
6. NAES buffer: 50 mM sodium acetate, pH 5.1, 10 mM EDTA, 1% sodium dodecyl sulfate (SDS).

2.2.2. Isolation of Eukaryotic RNA

1. Tissue was homogenized using FastRNA Tubes (BIO 101, no. 6040-601) and the BIO 101/Savant homogenizer FP120.
2. RNAzol (Biotecx CS 105).
3. mRNA isolation kit (Roche Molecular Biochemicals, no. 1 741 985), store at 4°C.
4. TE buffer: 1X 10 mM Tris-HCl, pH 7.5; 1 mM EDTA.
5. Tris-HCl buffer: 10 mM Tris-HCl, pH 8.0, 1 mM EDTA (Gibco-BRL).

2.3. RNA Labeling

2.3.1. Chemical Labeling of Total RNA

1. Psoralen-biotin (Schleicher & Schuell).
2. 350-nm ultraviolet (UV) light used for crosslinking Psoralen-Biotin.
3. LabelIT-Biotin for chemical labeling (PanVera).

4. 5X RNA fragmentation buffer: 200 m*M* Tris-acetate, pH 8.1, 500 m*M* KOAce, 150 m*M* MgOAce.

2.3.2. Enzymatic RNA Labeling

1. Biotinylated nucleotides Bio-11-CTP (25 nmol) (ENZO Diagnostics, no. 42818); store at –20°C.
2. Bio-16-UTP (25 nmol) (ENZO Diagnostics, no. 42814); store at –20°C.
3. Transcription kits (Superscript cDNA Synthesis Kit, no. 11594-017; Life Technologies); store at –20°C.
4. MEGAscript T7 Kit Ambion, no. 1334); store at –20°C (*see* **Note 2**).
5. T7-(T)$_{24}$ primer sequence: 5' GGCCAGTGAATTGTAATACGACTCACTATA GGGAGGCGG-(T)$_{24}$ VN 3' (*see* **Note 3**).
6. NH$_4$OAc, 7.5 *M* (Sigma, no. A 2706).
7. Phase Lock Gel™ tubes (5 Prime→3 Prime, Inc., pl-188233).
8. RNeasy Mini Kit (Qiagen, kit 74104); store at ambient temperature.
9. Phenol:chloroform:isoamyl alcohol mix (25:24:1) (Life Technologies, no. 15593-031); store at 4°C in the dark.
10. T7 10X ATP (75 m*M*) (Ambion).
11. T7 10X GTP (75 m*M*) (Ambion).
12. T7 10X CTP (75 m*M*) (Ambion).
13. T7 10X UTP (75 m*M*) (Ambion).
14. 10X T7 transcription buffer (Ambion).
15. 10X T7 enzyme mix (Ambion).

2.4. Hybridization and Antibody Signal Amplification

1. 1 L of 6X SSPE-T (pH 7.6): 300 mL of 20X SSPE (Gibco-BRL, no. 15591-043), 5 mL of 1% Triton X-100, 695 mL of dd-H$_2$O (*see* **Note 4**).
2. 1X MES buffer (pH 6.7): 0.1 *M* MES (Sigma, no. M 3023), 1.0 *M* NaCl, 0.01% Triton X-100.
3. 0.1X MES buffer (pH6.7): 0.1 *M* MES (Sigma, no. M 3023), 0.1 *M* NaCl, 0.01% Triton X-100.
4. 5X SAPE staining solution: 2 µL of streptavidin R–phycoerythrin (1 mg/mL) (Molecular Probes, cat. no. S-866), 10 µL of acetylated bovine serum albumin (aBSA) (20 mg/mL) (Sigma B, no. 8894), 188 µL of 1X MES buffer.
5. Biotinylated anti-SA staining solution: 1 µg/mL of biotinylated antistreptavidin (Vector, no. BA0500), and 0.5 mg/mL of aBSA (20 mg/mL) (Sigma B, no. 8894), in 1X of MES buffer.
6. 5X Fragmentation buffer: 40 mL of 1 *M* Tris-acetate, pH 8.1, 9.8 g of KOAc, 6.4 g of MgOAc; 140 mL of diethylpyrocarbonate (DEPC)-treated water; 200 mL of total volume.
7. Spike controls (BioB, BioC, BioD, Cre, checkerboard oligo B948) added to the hybridization solution are available from Affymetrix.
8. Herring sperm DNA stock (2 µL of 10 mg/mL) (Promega, no. D 1811).

2.5. Scanning and Data Analysis

Standard scanning conditions are predefined and included in the software for each array type. Individual analysis parameters are adjustable depending on the experiment. Data are analyzed with the expression data mining tool developed by Affymetrix. Details are available from the software manuals and help screens (*see* **Note 5**).

3. Methods
3.1. Isolation of Total RNA from Bacteria

This procedure for isolating total RNA from bacteria is derived from that of Maes and Messens (*9*). It yields 300–400 µg of RNA/50 mL of culture grown to an OD_{600} of 0.4. It is based on hot-phenol lysis.

1. Grow cells in 50 mL of the appropriate medium up to the desired optical density (OD).
2. Spin down the cells for 5 min at 8000*g*, discard the supernatant, and immediately freeze the cell pellet in liquid nitrogen.
3. Preheat phenol (previously equilibrated with NAE) at 60°C and add 3 mL of hot phenol to the frozen sample. After resuspension of the pellet and vortexing, incubate at 60°C for 5 min.
4. Add 3 mL of preheated NAES buffer, vortex, and incubate for an additional 5 min at 60°C.
5. Cool on ice and separate the phases by centrifugation (5 min, 3000*g*). Perform at least two additional phenol extractions until the interface is clean using Phase Lock Gel tubes.
6. Precipitate the last extraction in 0.3 *M* Na-acetate, pH 4.8, and 1 vol of isopropanol (*see* **Note 6**).
7. Wash the pellet in 70% EtOH and air-dry.
8. Resuspend the pellet in 348.5 µL of DEPC-treated water, 40 µL of RQ1 buffer, 10 µL of RNase-free DNase, 2 µL of RNasin, and incubate for 30 min at 37°C.
9. Add 5 µL of proteinase K and incubate at 37°C for 15 min. Phenol extract once or twice with Phase Lock Eppendorf tubes, and add Na-acetate (pH 4.8) to 0.3 *M*, and precipitate with 2.5 vol of EtOH.
10. Wash the pellet in 70% EtOH, air-dry the pellet, and resuspend in 30–50 µL of DEPC-treated water + 2 µL of RNasin. Then quantify total RNA by measuring of OD_{260}.

3.2. RNA Isolation from Tissues and Eukaryotic Cells

3.2.1. Tissues

1. Combine 50 mg of tissue and 1.0 mL of RNAzol in a FastRNA tube green.
2. Homogenize for 20 s at setting 6.
3. Add 0.1 vol of chloroform.

4. Homogenize 5 s at setting 6.
5. Transfer homogenate to a 2-mL Eppendorf tube.
6. Chill on ice for 5 min.
7. Spin at 12,000g for 15 min at 4°C (Eppendorf centrifuge 5417R; 2-mL tubes, 10,600 rpm).
8. Transfer the supernatant to a fresh 2-mL Eppendorf tube.
9. Add 2 vol of isopropanol and mix well.
10. Chill on ice for 15 min.
11. Spin at 12,000g for 10 min at 4°C (Eppendorf centrifuge 5417R; 2-mL tubes, 10,600 rpm).
12. Wash the pellet once with 1 mL of 75% EtOH.
13. Vortex and spin at 7500g for 5 min at 4°C (Eppendorf centrifuge 5417R; 1.5-mL tubes, 8500 rpm).
14. Dry briefly in a Speed-Vac (heater off).
15. Dissolve the RNA pellet in RNase-free water by pipetting and heating for 10 min at 55–60°C.
16. Check the aliquot on 1% formamide gel.

3.2.2. Cell Cultures

Cells are harvested and washed with an appropriate buffer prior to RNA isolation. In general, the cells should be kept at the culture temperature in order to avoid SOS-type responses, which do occur. For adherent lines, we recommend suspension of the cells in the culture vessel directly in RNAzol.

1. Resuspend the cell pellets (max. 108 cells) in 1 mL of RNAzol B each, and incubate on ice with occasional vortexing until lysis becomes apparent. For certain cell types, glass bead homogenization might be required for complete lysis.
2. Add 0.1 vol of chloroform.
3. Vortex for 15 s and chill for 5 min on ice.
4. Spin at 12,000g for 15 min at 4°C (Eppendorf centrifuge 5417R; 2-mL tubes, 10,600 rpm).
5. Transfer the aqueous phase to a fresh 2-mL tube and add 2 vol of isopropanol; mix well.
6. Incubate for 15 min on ice.
7. Spin at 12,000g for 10 min at 4°C (Eppendorf centrifuge 5417R; 2-mL tubes, 10,600 rpm).
8. Wash once with 1 mL 75% EtOH.
9. Vortex and spin at 7500g for 5 min at 4°C (Eppendorf centrifuge 5417R; 1.5-mL tubes, 8500 rpm).
10. Dry briefly in a Speed-Vac (heater off).
11. Dissolve the RNA pellet in RNase-free water by pipetting and heating for 10 min at 55–60°C.
12. Store frozen at –20°C.

3.2.3. Quantification of RNA

A Beckman DU series spectrophotometer allows use of microcells with a Micro-Auto-1 Accessory. These microcells need only 50 µL of solution and are ideal for using small (0.5–1 µL) aliquots of RNA solutions for quantification. Total RNA is dissolved in 20–400 µL of water depending on the expected amount and aiming for a 1 to 2 µg/µL concentration.

Fill the microcuvet with 50 µL of 1X TE for a blank reading. In the same cuvet, add 1 µL of RNA sample and mix with a micropipet. Record the absorbance at dual wavelengths of 260 and 280 nm. The following is an example of yield calculation:

1. $A260$ of 1 is equivalent to 40 µg of RNA/mL.
2. Therefore, take A260 × dilution factor × 40 to estimate the amount.
3. $A260 = 0.100$.
4. Dilution = 50 (1 µL of RNA in 50 µL of TE buffer in microcuvet).
5. Ext. coefficient = 40.
6. $0.1 \times 50 \times 40 = 200$ µg/mL or 0.2 µg/µL of the RNA solution.
7. Total RNA in solution = 10 µL.
8. Total RNA = 2 µg in 10 µL of undiluted sample.
9. For mRNA, the OD260/OD280 ratio should be close to 2.0.
10. RNA integrity should checked on a regular formamide gel prior to proceeding.

3.3. Labeling of Total Bacterial RNA

3.3.1. Psoralen-Biotin Labeling

1. Dilute total RNA (at least 50 mg) to about 200 ng/mL with DEPC-treated water, heat denature for 8 min at 95°C, and then quickly chill on ice water.
2. Add 7 nmol of psoralen biotin/50 µg of RNA (0.14 nmol/µg) and distribute the sample into a microtiter plate (no more than 150 mL/well). Place the microtiter plate under a 350-nm UV light for 3 h in a cold room to avoid overheating the sample.
3. Perform three water-saturated n-butanol extractions (2 vol), ethanol precipitate the sample (0.3 M NaAce), and resuspend into 12 µL of DEPC-treated water. Labeled material can be stored at –80°C for several weeks.
4. Perform fragmentation immediately before hybridization in fragmentation buffer for 40 min at 95°C. Fragmentation is performed in the smallest possible volume (e.g., 3 µL of fragmentation buffer in 12 µL). Quickly chill on ice, centrifuge, and prepare hybridization mix.

3.3.2. LabelIT-Biotin Labeling

LabelIT from PanVera was also successfully used to label total RNA. The chemical reaction involved in labeling is undisclosed. The reaction is most efficient at an RNA concentration of ±0.1 µg/µL. The lyophilized labeling reagent is first reconstituted in 100 µL dimethyl sulfoxide (DMSO). The best signals on chips are obtained with five times the amount of LabelIT reagent recommended by the manufacturer.

A typical labeling reaction for 50 µg of RNA contains 50 µL of 10X LabelIT buffer, 250 µL of LabelIT reagent, 50 µg of RNA in a final volume of 500 µL. The reaction is then incubated at 37°C for 2 h. Under these conditions, chip hybridization signals are four to five times more intense when compared to psoralen biotin labeling.

3.4. Eukaryotic mRNA Labeling

3.4.1. cDNA Synthesis

From poly A+ mRNA. All reagents are part of the Life Technologies custom SuperScript choice GeneChip kit (cat. no. 11594-017) (*see* **Fig. 2**).

3.4.1.1. FIRST-STRAND SYNTHESIS

1. Add the following to a 1.5-mL tube: 0.1–5 µg of poly A$^+$ mRNA (approx 1 µg/µL), 1 µL of T7-(T)24 Primer (100 pmol/µL), and DEPC-H2O to a total volume of 12 mL.
2. Incubate at 70°C for 10 min.
3. Chill on ice.
4. Add the following components (on ice): 4 µL of 5X first-strand buffer, 2 µL of 0.1 *M* DTT, and 1 µL dNTP mix (10 m*M*).
5. Incubate for 2 min at 37°C.
6. Add 1 µL of SSII RT/µg RNA. For less than 1 µg of RNA, use 1 µL of SSII RT. Mix well and spin briefly. The final volume should be 20 µL.
7. Incubate for 1 h at 37°C

3.4.1.2. SECOND-STRAND SYNTHESIS

1. Spin first-strand cDNA reactions briefly and place them on ice.
2. Add 91 µL of DEPC-H2O, 30 µL of 5X second-strand buffer, 3 µL of dNTP mix (10 m*M*), 1 µL of *E. coli* DNA Ligase (10 U/µL), 4 µL of *E. coli* DNA Polymerase I (10 U/µL), and 1 µL of RNase H (2 U/µL).
3. Mix well, spin briefly, and incubate at 16°C for 2 h.
4. Add 2 µL of T4 DNA Polymerase (10 U).
5. Incubate at 16°C for 5 min.

3.4.1.3. cDNA CLEANUP USING PHENOL-CHLOROFORM EXTRACTION AND PHASE LOCK GEL

1. Add an equal volume (162 µL) of 25:24:1 phenol:chloroform:isoamyl alcohol (saturated with 10 m*M* Tris-HCl) to the cDNA sample and vortex.
2. Pellet a Phase Lock Gel (1.5-mL tube) for 20–30 s.
3. Transfer the entire cDNA-phenol/chloroform mixture to the phase lock gel tube.
4. Spin at full speed (12,000*g* or higher) for 2 min.
5. Transfer the upper aqueous phase to a fresh RNase-free 1.5-mL tube.

6. Add 0.5 vol of 7.5 M NH4OAc and 2.5 vol of ice-cold 100% ethanol and vortex (2.5 vol of aqueous phase plus ammonium acetate volume).
7. Immediately centrifuge at 14,000–16,000g at room temperature for 20 min.
8. Wash the pellet carefully two times with 500 µL of cold (–20°C) 80% ethanol.
9. Dry the pellet 5–10 min in the Speed-Vac.
10. Dissolve the cDNA in a small volume of DEPC-treated water (typical yield is 0.25–0.65 µg/µL of cDNA; the maximum volume that can be added to the following in vitro transcription [IVT] reaction is 1.5 µL).
11. Check an aliquot (one-tenth of the sample) on a 1% TBE-agarose gel.
12. Proceed or store the sample at –20°C.

3.4.2. Direct cDNA Synthesis from Total RNA (Direct cDNA)

3.4.2.1. FIRST-STRAND SYNTHESIS

1. Add the following to a 1.5-mL tube: 20 µg of total RNA, 2 µL of T7-(T)24 Primer (100 pmol/µL), DEPC-water to 22 µL.
2. Incubate at 70°C for 10 min.
3. Chill on ice.
4. Add the following components (on ice): 8 µL of 5X first-strand buffer, 4 µL of 0.1 M DTT, 2 µL of dNTP mix (10 mM).
5. Mix well, spin briefly, and incubate for 2 min at 37°C.
6. Add 4 µL of SSII RT, mix well, and spin briefly.
7. Incubate for 1 h at 37°C.

3.4.2.2. SECOND-STRAND SYNTHESIS

1. Spin first-strand cDNA reactions briefly and place them on ice.
2. Add 182 µL of DEPC-water, 60 µL of 5X second-strand buffer, 6 µL of dNTP mix (10 mM), 2 µL of *E. coli* DNA Ligase (10 U/µL), 8 µL of *E. coli* DNA Polymerase I (10 U/µL), 2 µL of RNase H (2 U/µL).
3. Mix well, spin briefly, and incubate at 16°C for 2 h.
4. Add 4 µL of T4 DNA Polymerase (5 U/µL).
5. Incubate at 16°C for 5 min.
6. Add 0.5 µL of RNase A (100 µg/µL).
7. Incubate at 37°C for 30 min.
8. Add 7.5 µL of Proteinase K (10 µg/µL) and 7.5 µL of 20% SDS.
9. Incubate at 37°C for 30 min.
10. Add 20 µL of 0.5 M EDTA and store at –20°C or proceed with the phase lock gel cleanup procedure as described above.

3.4.3. IVT Labeling of cDNA

3.4.3.1. IVT REACTION

1. Prepare NTP labeling mix for four IVT-reactions (store unused labeling mix at –20°C): 8 µL of T7 10X ATP (75 mM), 8 µL of T7 10X GTP (75 mM), 6 µL of T7 10X CTP (75 mM), 6 µL of T7 10X UTP (75 mM), 15 µL of Bio-11-CTP (10 mM), 15 µL of Bio-16-UTP (10 mM).

2. For each reaction combine the following on RT (not on ice): 14.5 µL of NTP labeling mix, 2.0 µL of 10X T7 transcription buffer, 1.5 µL of ds cDNA (0.3–1 µg is optimal) (*see* **Note 7**), 2.0 µL of 10X T7 enzyme mix.
3. Incubate 4–6 h at 37°C.
4. Purify the sample immediately after IVT. Freezing is not recommended because stable RNA secondary structures can potentially be formed.

3.4.3.2. IVT CLEANUP USING RNEASY SPIN COLUMNS FROM (*SEE* NOTES 8–12)

1. Take half of your IVT and adjust the sample volume to 100 µL with RNase-free water.
2. Add 350 µL of Buffer RLT to the sample and mix thoroughly.
3. Add 250 µL of ethanol (96–100%) and mix well by pipetting; do not centrifuge.
4. Apply the sample (700 µL) to an RNeasy mini spin column sitting in a collection tube.
5. Centrifuge for 15 s at ≥8000g (Eppendorf centrifuge 5417C; 11,000 rpm).
6. Transfer the RNeasy column into a new 2-mL collection tube.
7. Add 500 µL of Buffer RPE and centrifuge for 15 s at ≥8000g (Eppendorf centrifuge 5417C; 11,000 rpm); discard flowthrough and reuse the collection tube.
8. Pipet 500 µL of RPE Buffer onto the RNeasy column and centrifuge for 2 min at maximum speed to dry the RNeasy membrane.
9. Transfer the RNeasy column into a new 1.5-mL collection tube (supplied) and pipet 30 µL of RNase-free water directly onto the RNeasy membrane; centrifuge for 1 min at ≥8000g (Eppendorf centrifuge 5417C; 11,000 rpm) to elute.
10. Repeat **step 8**; elute into the same collection tube using another 30 µL of RNase-free water.
11. Check 1–10% of the sample on a formaldehyde gel and measure the OD.

3.4.4. Fragmentation of the IVT Products (see **Note 13**)

1. Mix 16 µL cRNA with 4 µL of 5X fragmentation buffer.
2. Incubate at 95°C for 35 min. Place on ice after incubation.
3. Store fragmented cRNA at –20°C or proceed with the hybridization.

3.5. Hybridization of the Chip

Sample composition (200 µL) contains the following: 10 µg of fragmented, labeled cRNA; 0.1 mg/mL of HS DNA; 0.5 mg/mL of acetylated BSA (2 mL of 50 mg/mL or 6.25 mL of 20 mg/mL stock); 2 µL of spiked controls; 2 µL of Oligo B948 (5 nM); and 1X MES buffer to a 200 uL volume.

1. Apply 200 µL of chip pretreatment solution (0.5 mg/mL of acetylated BSA, 0.5 mg/mL of HS DNA in 1X MES buffer) to the chip and incubate at 40°C for 15 min with rotation.
2. Rinse the chip briefly with 1X MES buffer.
3. Heat the sample at 99°C for 5 min and then put in a 45°C water bath for 10 min before applying to the chip.

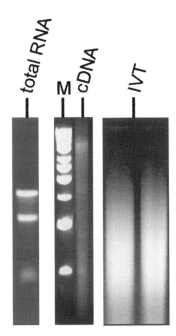

Fig. 2. Typical nucleic acid migration patterns during different stages of sample preparation. All samples are from mouse brain. Note that the cDNA smear goes from the top to the bottom of the gel whereas the in vitro transcripts (IVT) are typically smaller. M, indicates DNA size markers.

4. Add the sample to the chip and hybridize overnight at 45°C with rotation (Heidolph REAX-2; speed setting 7).
5. Wash on a fluidics station with the WASH-A program provided.
6. Rinse the chip with 0.1X MES buffer and then apply 200 µL of fresh 0.1X MES buffer to wash the chip at 45°C for 30 min with rotation.
7. Rinse the chip with 1X MES buffer.
8. Stain with 5X SAPE staining solution at 40°C for 15 min with rotation.
9. Wash on the fluidics station with the WASH-A program.
10. Scan the chip or proceed with antibody amplification.

3.6. Biotinylated Anti-SA Stain

1. Remove 6X SSPE-T solution inside the chip cartridge and rinse the chip with 200 µL of 1X MES buffer.
2. Stain the chip with 200 µL of Anti-SA staining solution and incubate at 40°C for 30 min with rotation.
3. Rinse the chip a few times with 200 µL of 1X MES.
4. Wash the chip on the fluidics station with the WASH-A program.

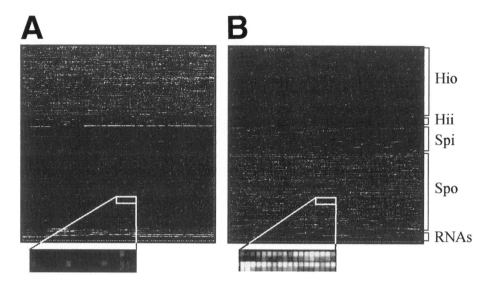

Fig. 3. Fluorescence image of an array with 250,000 different oligonucleotide probes covering 4870 sequences from *H. influenzae* and *S. pneumoniae*. The image was obtained following hybridization of biotin-labeled randomly fragmented RNA from *H. influenzae* (**A**) and from *S. pneumoniae* (**B**). Enlargement of a small area of the chip is also shown at the bottom. Hio, probes covering *H. influenzae* open reading frames; Hii, probes covering *H. influenzae* intergenic regions larger than 200 bp; Spo, probes covering *S. pneumoniae* open reading frames; Spi, probes covering *S. pneumoniae* intergenic regions larger than 200 bp; RNAs, probes covering nonunique ribosomal and transfer RNA sequences.

5. Use 1X MES buffer to rinse the chip.
6. Apply 200 μL of 5X SAPE stain solution to stain the chip and incubate at 40°C for 15 min with rotation.
7. Rinse the chip briefly with 1X MES buffer.
8. Wash the chip on the fluidics station with the WASH-A program.

3.7. Scanning and Data Analysis

We use an Affymetrix-Hewlett-Packard Scanner at 570 nm for SAPE-stained chips: pixel size setting 3 μm for a 24-μm chip, or 9 μm for 50-μm chip. Computation of the relative expression level of all genes is performed by the Affymetrix GeneChip software (*6*). The transcript level of a gene is determined based on the probe set intensity and is given by the average of the differences (perfect match minus mismatch hybridization signal) for each gene. **Figure 3** gives an example of a scanned bacterial chip hybridized with chemically biotinylated RNA from *H. influenzae* or *S. pneumoniae*.

4. Notes

1. Currently, we are working on random reverse transcription of total bacterial RNA based on the method developed by Chuang et al. *(10)*. In such experiments, antisense chips are used for hybridization of cDNA. For information about the technology, we recommend Certa et al. *(2)* and de Saizieu et al. *(11)*.
2. The cDNA kit is especially adapted for preparing chip probes and may not be available in all countries. Therefore, we recommend the regular kit, which is more expensive and does not include the appropriate primer (no. 18090-019).
3. It is essential that the primer is polymerase gel electrophoresis-gel purified and of high quality (Microsynth GmbH, Schützenstrasse 15, 9436 Balgach, Switzerland; phone: ++41 71 722 83 33; fax: ++41 71 722 87 58; e-mail: oligo@microsynth.ch).
4. Buffers are filter sterilized and stable at room temperature for 3 mo. The correct pH of 6.7 is important. The staining and antibody amplification solutions are made up fresh prior to use. The aBSA and the antibody are stored frozen and phycoerythrin is stored in aliquots at 4°C. Never freeze the phycoerythrin because this will result in an almost complete loss of activity (>90%). Always keep a backup tube of phycoerythrin in a separate refrigerator, because integrity and availability of this reagent is critical for chip experiments in general.
5. Some users have developed their own data analysis packages, and professional software developers provide solutions to analyze the complex chip data. A scan of a high-density chip generates about 70 megabytes of data. Storage and handling of data is therefore an important issue.
6. Precipitation with ethanol may lead to more impurities in the final preparation.
7. Do not add more than 1 µg of cDNA; higher concentrations can inhibit the IVT reaction, which results in lower yields.
8. Save an aliquot of the unpurified IVT product to analyze by gel electrophoresis.
9. We suggest purifying half of the IVT reaction and checking yields before purifying the second half. Occasionally, the sample is lost during the purification, but the remaining half is still available. The total amount of RNA in your IVT reaction may exceed the capacity of the matrix used for purification. Therefore, better overall yields will be obtained by purifying half of the reaction at a time.
10. It is quite important to purify the samples immediately after IVT. If you have to purify your sample after freezing, be sure to incubate it at 65°C for 10–15 min before cleanup.
11. Pass the sample over the column twice before the wash and elution steps. After adding water to the column for elution, wait 1 min before centrifuging.
12. review the Qiagen manual for additional information.
13. Typically, cRNA is hydrolyzed at 1 µg/µL in 20 µL of buffer or less.

Acknowledgments

We thank Monika Seiler and Karin Kuratli for substantial contributions to the technology and protocol optimization. We also thank David Lockhart for communication of the MES-hybridization atlantic-protocol prior to the official release, and Janet Warrington and Archana Nair for support and help in the initial development phase of the technology.

References

1. Schena, M., Shalon, D., Davis, R. W., and Brown, P. O. (1995) Quantitative monitoring of gene expression patterns with a complementary DNA microarray. *Science* **270,** 467–470.
2. Certa, U., Hochstrasser, R., Langen, H., Buess, M., and Maroni, C. (1999) Biosensors in biomedical research: Development and applications of gene chips. *Chimia* **53,** 57–61.
3. Fodor, S. P. A., Read, J. L., Pirrung, M. C., Stryer, L., Lu, A. T., and Solas, T. (1991) Light-directed spatially addressable parallel chemical synthesis. *Science* **251,** 767–773.
4. Fodor, S. P., Rava, R. P., Huang, X. C., Pease, A. C., Holmes, C. P., and Adams, C. L. (1993) Multiplexed biochemical assays with biological chips. *Nature* **364,** 555–556.
5. Pease, A. C., Solas, D., Sullivan, E. J., Cronin, M. T., Holmes, C. P., and Fodor, S. P. A. (1994) Light-directed oligonucleotide arrays for rapid DNA sequence analysis. *Proc. Natl. Acad. Sci. USA* **91,** 5022–5026.
6. Lockhart, D. J., Dong, H., Byrne, M. C., Follettie, M. T., Gallo, M. V., Chee, M. S., Mittmann, M., et al. (1996) Expression monitoring by hybridization to high-density oligonucleotide arrays. *Nat. Biotechnol.* **14,** 1675–1680.
7. Wodicka, L., Dong, H., Mittmann, M., Ho, M. H., and Lockhart, D. J. (1997) Genome-wide expression monitoring in Saccharomyces cerevisiae. *Nat. Biotechnol.* **15,** 1359–1367.
8. de Saizieu, A., Certa, U., Warrington, J., Gray, C., Keck, W., and Mous, J. (1998) Bacterial transcript imaging by hybridization of total RNA to oligonucleotide arrays. *Nat. Biotechnol.* **16,** 45–48.
9. Maes, M. and Messens, E. (1992) Phenol as a grinding material in RNA preparations. *Nucleic Acids Res.* **20,** 4374.
10. Chuang, S.-E., Daniels, D. L., and Blattner, F. R. (1993) Global regulation of gene expression in *Escherichia coli. J. Bacteriol.* **175,** 2026–2036.
11. de Saizieu, A., Gardès, C., Flint, N., Wagner, C., Kamber, M., Mitchell, T. J., et al. (2000) Microarray based identification of a novel *Streptococcus* pneumoniae regulon controlled by an autoinduced-peptide. *J. Bacteriol.* **182,** 4696–4703.

10

Analysis of Nucleic Acids by Tandem Hybridization on Oligonucleotide Microarrays

Rogelio Maldonado-Rodriguez and Kenneth L. Beattie

1. Introduction

Oligonucleotide arrays (also known as DNA chips, gene chips, or genosensor chips) are emerging as a powerful research tool in nucleic acid sequence analysis. Several technical challenges remain to be solved, however, before oligonucleotide arrays can reach their full potential and be implemented in a robust fashion. Some of these challenges are as follows:

1. The need to generate single-stranded target nucleic acids in order to achieve optimal hybridization signals.
2. Spontaneous formation of secondary structure in the single-stranded target nucleic acid, causing certain stretches of target sequence to be poorly accessible to hybridization.
3. Imperfect specificity of hybridization, making it difficult or impossible to distinguish between certain sequence variations.
4. The strong influence of base composition on the stability of short duplex structures, making it difficult to use an extensive array of oligonucleotides (differing in base composition) to analyze a nucleic acid sample under a single hybridization condition.
5. Multiple occurrence of sequences complementary to short oligonucleotide probes within the nucleic acid sample, limiting the genetic complexity of a nucleic acid sample that can be analyzed by arrays of short oligonucleotide probes.
6. The need to label each nucleic acid analyte prior to hybridization to the DNA probe array, a significant factor in the overall time and cost of analysis.

The tandem hybridization strategy described in this chapter is designed to minimize or eliminate these difficulties.

From: *Methods in Molecular Biology, vol. 170: DNA Arrays: Methods and Protocols*
Edited by: J. B. Rampal © Humana Press Inc., Totowa, NJ

Stabilization of short duplex structures by base stacking interactions between tandemly hybridized (contiguously stacked) oligonucleotides has been reported by several laboratories *(1–10)*. The Mirzabekov group proposed that the stabilizing effect of contiguous stacking could be used to improve the efficiency of *de novo* DNA sequencing by hybridization *(1,3,4)* and to detect point mutations *(2)*. We recently reported a novel tandem hybridization strategy *(11,12)* that differs in several respects from the contiguous stacking hybridization approach described by the Mirzabekov laboratory (*see* **Note 1** for a discussion of the differences). **Figure 1** illustrates the tandem hybridization procedure. The nucleic acid sample is first annealed with a molar excess of one or more labeled "stacking probes," typically 20–30 bases in length, each complementary to a unique target sequence within the nucleic acid analyte. The mixture is then hybridized to an array of short surface-tethered "capture probes" (typically 7- to 9-mer), each designed to hybridize in tandem with a stacking probe on the target sequence of interest. Hybridization is carried out at a temperature whereby the short capture probes will not form stable duplex structures with the target sequence unless they form contiguous base stacking with their corresponding longer labeled probes (*see* **Fig. 1**). The quantitative hybridization fingerprint reveals the occurrence and relative abundance of cognate sequences within the nucleic acid sample.

The tandem hybridization method described herein offers several advantages over the traditional oligonucleotide array configuration in which each interrogated target sequence is represented by a single surface-tethered probe.

First, the stacking probe (preannealed to the nucleic acid sample) contributes the label to the target sequence, eliminating the step of separately labeling each sample prior to analysis. This can significantly reduce analysis time and cost when large numbers of samples need to be analyzed. Furthermore, problems of nonspecific hybridization may be reduced when a complex nucleic acid is being analyzed, because the label is incorporated specifically into the nucleic acid fragment(s) of interest.

Second, because the long stacking probe targets the analysis to a single unique site within a nucleic acid, direct hybridization analysis of nucleic acid samples of high genetic complexity using short capture probes is enabled, which can eliminate the need to prepare specific target fragments using polymerase chain reaction (PCR).

Third, the stacking probe (which can be positioned on one or both sides of the capture probe) can eliminate interfering secondary structures within the target strand, such as stem-loop (hairpin) structures involving the target sequence of interest. In addition, preannealing of stacking probes (in molar excess) to heat-denatured duplex DNA can eliminate the need to prepare single-stranded target DNA prior to hybridization.

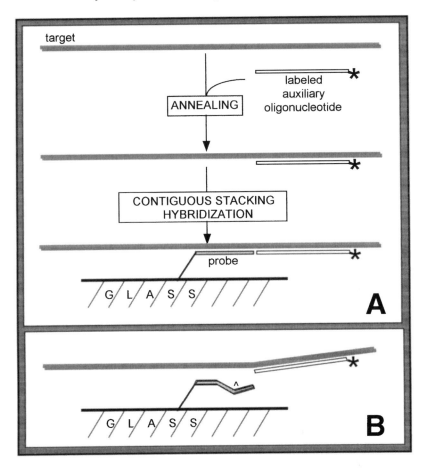

Fig. 1. Tandem hybridization strategy. As shown in (**A**), a nucleic acid target is first annealed with a molar excess of labeled auxiliary oligonucleotide, typically 20–30 bases long, designed to bind to a unique site within the nucleic acid sample. The labeled auxiliary oligonucleotide (the stacking probe) serves as a reagent to introduce the detection tag into a single specific site within the nucleic acid target. The partially duplex labled target is then applied to the array of capture probes, typically 7–12 bases long, end-tethered to a flass slide or chip. Each capture probe is designed to hybridize with the target strand in tandem with a labeled stacking probe preannealed to the target. Base stacking interactions between the short capture probe and the long stacking probe stabilize the binding of partially duplex labeled target to the glass-tethered capture probe. Hybridization is carried out at an (elevated) temperature at which the short capture probe by itself does not form a stable duplex structure with the target; capture of label to the glass occurs only when the capture probe hybridizes in tandem with the labeled stacking probe. As depicted in (**B**), a single base mismatch between the short capture probe and the target sequence will disrupt the short duplex, preventing binding of label to the glass.

Fourth, because hybridization of the short capture probe is strengthened by base stacking interactions propagated through the long stacking probe, base composition effects within the short capture probes may be minimized, facilitating the simultaneous use of a large array of capture probes under a single hybridization condition.

Fifth, excellent hybridization specificity is achieved: all base mismatches at internal positions within the short capture probe and most terminal mismatches at the junction of capture and stacking probes are efficiently discriminated against, facilitating analysis of sequence variations (mutations and DNA sequence polymorphisms).

The tandem hybridization strategy offers improved performance in many important oligonucleotide hybridization applications, including the following:

1. Repetitive analysis of known mutations or sequence polymorphisms in numerous genomic samples.
2. Simultaneous analysis of numerous known mutations or sequence polymorphisms in single genes, a multiplicity of genes, or on a genomewide scale.
3. Identification of species, strains, or individuals through the use of oligonucleotide probes and auxiliary oligonucleotides targeted to nucleotide sequences known to be unique for such species, strains or individuals.
4. Analysis of gene expression (transcription) profiles through the use of oligonucleotide probes and auxiliary oligonucleotides known to be unique to individual mRNA species.
5. Analysis of nucleic acid samples of high genetic complexity.

Furthermore, the strategy of labeling the target strands by annealing with labeled auxiliary oligonucleotides can increase the convenience and cost-effectiveness of high-throughput nucleic acid sequence analyses, and the labeling can be done with a number of nonradioactive tags, using reagents that generate fluorescence, chemiluminescence, or visible colors in the detection step.

This chapter is intended as a general guide for implementing the tandem hybridization approach in a simple, robust format, accessible to any laboratory. Incremental improvements in performance will inevitably occur as the methods are applied in other laboratories and modifications are made in aspects such as attachment chemistry, hybridization supports, and detection schemes. Of course, ultrahigh sample throughput and genomewide scale of analysis may be achieved with the availability of sophisticated, expensive instrumentation and further development in technology.

For examples of experimental results obtained using tandem hybridization approaches described herein and by the Mirzabekov laboratory, the reader is referred to **refs.** *1*, *2*, *4*, *11*, and *12*. Differences between the tandem hybridization procedures described by the Maldonado-Rodriguez and Mirzabekov laboratories are discussed in **Note 1**.

2. Materials

Unless indicated otherwise, all chemicals should be analytical reagent grade or higher.

1. Hexane.
2. Ethanol, absolute.

2.1. Oligonucleotide Arrays

The tandem hybridization strategy described herein may be used with any type of oligonucleotide arrays, independent of the immobilization chemistry and type of hybridization support.

1. Microscope slides: Any brand of microscope slides will suffice, provided that they are cleaned using the simple protocol provided (*see* **Subheading 3.1.**).
2. Micropipeting device capable of reasonably precise delivery of 200-nL vol: For manual spotting of arrays, a properly calibrated Gilson P-2 micropipettor is recommended.
3. Synthetic oligonucleotides, derivatized at the 3' end with C3-amine (*see* **Note 2**).
4. 3'-Amino-Modifier C3 CPG support from Glen Research (Sterling, VA).
5. 3'-Amine-ON™ CPG from Clonetech (Palo Alto, CA).

2.2. Target Nucleic Acid

The target strands can be PCR products, RNA, genomic DNA, or synthetic oligonucleotides (*see* **Note 3**). If PCR is used to amplify the target sequence, it is recommended that the PCR product be converted to single-stranded form to give optimal hybridization. For this purpose, one of the two PCR primers is labeled with biotin on the 5' terminus by use of biotin phosphoramidite during the final coupling step in the chemical synthesis. The following materials are needed to isolate the nonbiotinylated target strand prior to hybridization:

1. Ultrafree-MC Filters (30,000 NMWL regenerated cellulose) from Millipore (Bedford, MA).
2. AffiniTip streptavidin capture microcolumns from Sigma Genosys (The Woodlands, TX).
3. 1X AffiniTip binding buffer: 5 mM Tris-HCl (pH 7.5), 1 M NaCl, 0.5 mM EDTA, 0.05% Tween-20.

2.3. Hybridization

1. 20X Standard saline citrate (SSC): 3 M NaCl, 0.3 M sodium citrate, pH 7.0 from 5PRIME->3PRIME (Boulder, CO).
2. Tetramethylammonium chloride (TMAC).
3. EDTA.
4. Tris(hydroxymethyl)aminomethane (Tris), ultraPURE grade from Life Technologies (Gaithersburg, MD).

5. Sodium dodecyl sulfate (SDS).
6. Tripolyphosphate.
7. Polyethylene glycol 8000.
8. Hybridization chamber: small sealable plastic box (e.g., a micropipet tip box) containing a shelf for holding the slides, a water reservoir at the bottom, and a thick layer of water-saturated filter paper at the top.
9. Hybridization buffer: 3.3 M TMAC dissolved in 50 mM Tris-HCl, pH 8.0, 2 mM EDTA, 0.1% (w/v) SDS and 10% (w/v) polyethylene glycol.
10. Glass or plastic cover slips.
11. [γ-^{32}P]ATP, 7000 Ci/mmol.

2.4. Detection/Image Analysis

Numerous types of detection and imaging systems can be used for analyzing hybridization patterns obtained using oligonucleotide arrays. The major options are listed next, but a detailed description of their use is beyond the scope of this chapter.

1. X-ray film and flatbed scanner.
2. Phosphorimager (Molecular Dynamics, Bio-Rad, or Fuji).
3. Charge-coupled device (CCD) imaging system.

3. Methods
3.1. Oligonucleotide Arrays

A basic protocol is given here for 3'-end attachment of oligonucleotides to glass slides. This procedure is faster and more convenient than the 5'-amine/ epoxysilane attachment method reported previously *(13,14)*, requiring only cleaning but no chemical derivatization of the glass surface *(15–18)*.

1. Synthesize oligonucleotides by means of the standard phosphoramidite procedure *(19)* using the 3'-Amino-Modifier C3 CPG support or the 3'-Amine-ON™ CPG (*see* **Note 2**).
2. To clean the glass slides prior to attachment of oligonucleotides, soak slides in hexane for 15 min, rinse briefly with ethanol, and then dry at 80°C for at least 1 h.
3. Dissolve oligonucleotide probes of desired sequence and length, containing 3'-aminopropanol modification, in sterile deionized H$_2$O at a final concentration of 20 μM, and manually apply 200-nL droplets of each probe to the clean glass slides using a Gilson P-2 Pipetman or, if available, using a robotic fluid-dispensing system. Rows of three droplets of each probe can be attached to improve the reproducibility of the results. Allow droplets to dry (approx 15 min at room temperature). Then rinse slides in water and use immediately in the prehybridization step or, if necessary, air-dry and store in a vacuum desiccator.
4. Just before hybridization, soak the slides for 1 h at room temperature with blocking agent (10 mM tripolyphosphate) and then rinse with water and use immediately for hybridization (*see* **Note 4**).

3.2. Target Nucleic Acid: Single-Stranded PCR Products Annealed with Labeled Stacking Oligonucleotides

1. Synthesize the PCR primer complementary to the target strand with a 5'-biotin modification.
2. To remove the excess of biotinylated primers from the PCR reaction, apply approx 50 ng or 200–500 fmol of a PCR product in 50 µL (*see* **Note 5**) to the sample cup of a Millipore Ultrafree spin filter (30,000 *Mr* cutoff), followed by 350 µL of sterile deionized water, and centrifuge the spin filter at 2000*g* for 5 min at room temperature in a microfuge, until the volume is reduced to 5–10 µL. Add an additional 250 µL of sterile deionized water and centrifuge the filter again for 5 min until the volume is reduced to 1–5 µL. Suspend the "retained" material in 50 µL of sterile 1X AffiniTip binding buffer and slowly draw up into an AffiniTip™ column attached to a Gilson P200 Pipetman. Slowly expel the material and draw into the AffiniTip ten times. Then add another 50 µL of 1X AffiniTip binding buffer with mixing and draw the second suspension into the AffiniTip and slowly aspirate/expel the 100-µL vol 10 more times. After incubation for 15 min in the AffiniTip, again expell/aspirate the liquid 5–10 times. Then completely expel the liquid and wash the AffiniTip by aspirating/expelling five 200-µL aliquots of 1X binding buffer, followed by one 1000-µL vol of 1X binding buffer, and, finally, 1000 µL of sterile deionized water. After all the water is expelled, elute the nonbiotinylated single-stranded DNA fragment with 20 µL of 0.2 *N* NaOH (aspirate/expel 10 times, incubate 10 min, then aspirate/expel another 10 times), collect in a microfuge tube and neutralize with 4 µL of 1 *N* HCl.
3. Formation of partially duplex, labeled target DNA molecules: Preanneal ^{32}P-labeled stacking oligonucleotides, preferably 15-mer or longer (*see* **Note 6**), to the single-stranded PCR fragments, as follows. Label 5 pmol of each stacking oligonucleotide by kinasing with an excess of γ-^{32}P-ATP (23 m*M*, specific activity 7000 Ci/mmol). Anneal aliquots of single-stranded target DNA with one or more labeled auxiliary oligonucleotides. The annealing mixture contains 50 µL of 20X SSC, 10 µL of 1 *M* Tris-HCl (pH 8.0), 3 µL of 0.5 *M* EDTA, one or more prelabeled auxiliary oligonucleotide (0.2 pmol each), 10 µL of single-stranded target DNA (typically 0.1 pmol), and high-performance liquid chromatography (HPLC)-pure H_2O to 90 µL. Incubate the mixture at 95°C for 5 min, 45°C for 5 min, then 6°C for 5 min. Remove excess γ-^{32}P[ATP] by microcentrifugation through an Ultrafree spin filter (30,000 *M$_r$* cutoff), and dissolve the retained DNA in 20 µL of 1X SSC.

3.3. Hybridization

3.3.1. General Protocol

The following basic protocol is applicable to analysis of single-stranded PCR products. As discussed in subsequent sections, the general protocol can be modified for use with other types of nucleic acid samples.

1. Carry out hybridization of single-stranded PCR products to oligonucleotide arrays tethered to glass slides in hybridization buffer. If desired, TMAC may be replaced with 6X SSC (*see* **Note 7**).
2. Place the hybridization mixture containing target DNA, preannealed with one or more labeled stacking oligonucleotides (20-μL vol), onto each slide and apply a cover slip.
3. Place the slides in a hybridization chamber (*see* **Subheading 2.3., item 8**) and incubate overnight at 15°C or a higher temperature, depending on the length of the capture probe (*see* **Note 8**).
4. After hybridization, wash the slides by dipping several times in the corresponding hybridization buffer without polyethylene glycol, air-dry, and then wrap in plastic film and place against X-ray film for autoradiography, or image using a phosphorimager or CCD imaging system. Multiple sites within a single PCR product, or individual sites within multiple PCR products, can be analyzed simultaneously if the PCR product is preannealed with multiple labeled stacking probes (each designed to anneal adjacent to a specific glass-tethered capture probe) prior to hybridization to the array of capture probes on the slide.

3.3.2. Analysis of Double-Stranded PCR Products

The procedure for analyzing double-stranded PCR products is essentially the same as described in the previous section, except that a double-stranded PCR product is heat denatured and preannealed with a molar excess of two stacking oligonucleotides complementary to the target strand, which are designed to anneal to the target strand producing a single-stranded gap equivalent to the length of the glass-tethered capture probe. In other words, there are two contiguously stacking probes, one on each side of the capture probe, at least one of which is labeled. For simultaneous analysis of multiple sites within a PCR product or individual regions within multiple PCR fragments, multiple pairs of stacking oligonucleotides can be simultaneously annealed to the PCR product(s). Once annealed, the excess of stacking olgonucleotides can be removed by microfiltration through Ultrafree spin-filters as described (*see* **Subheading 3.2.**) and the retained DNA recovered in hybridization buffer, ready for hybridization to the array of capture probes on the glass slide.

3.3.3. Analysis of Nucleic Acids of High Genetic Complexity

The foregoing protocols are used when amplification of target strands by PCR or other target enhancement methods is required. In many types of nucleic acid analysis, such as microbial identification and gene expression/mRNA profiling, the quantity of nucleic acid will be sufficient to enable direct analysis using the tandem hybridization strategy, without the need to perform DNA amplification.

The critical consideration that enables direct application of the tandem hybridization approach without DNA amplification is the appropriate design of the labeled stacking probes, ensuring that the hybridization is specifically targeted to unique sites within nucleic acid analytes of high genetic complexity. To achieve the required site specificity with a nucleic acid sample such as total genomic DNA or bulk mRNA (or cDNA derived therefrom), the labeled stacking probes are selected of sufficient length (typically 15–30 bases, the exact length depending on the genetic complexity of the nucleic acid analyte) to ensure that each stacking probe binds to a unique position within the nucleic acid analyte. *See* **Note 6** for guidance concerning the appropriate length of stacking probe to use with a nucleic acid sample of a given genetic complexity.

The labeled stacking probes are added to the nucleic acid sample, and the mixture is hybridized with the array of short capture probes (typically 7–10 bases in length), each designed to bind to a specific target sequence, in tandem with the corresponding longer stacking probe. The capture probes may be designed to bind to the target on either the 5' or 3' side of the tandemly hybridizing stacking probe. The labeled stacking probes are typically preannealed to the nucleic acid analyte but, alternatively, may be added to the analyte at the beginning of hybridization to the oligonucleotide array, or even after application of analyte to the array. The hybridization is carried out under conditions such that significant binding of the nucleic acid analyte to the array of capture probes will occur only if the capture probe and stacking probe hybridize in tandem with the target strand, so that base stacking interactions at the junction of the stacking and capture probes stabilize the binding. For example, under typical hybridization conditions (6X SSC or 3 *M* TMAC at 45°C) a 9-mer capture probe will not by itself (i.e., in the absence of tandemly hybridizing stacking probe) form a stable duplex structure with complementary sequences within a nucleic acid sample. For a nucleic acid analyte of high genetic complexity, a given short capture probe will likely have numerous complements (*see* **Note 6**); however, the long stacking probe serves to direct the binding to the single site of interest within the nucleic acid. The optimal hybridization temperature and selection of capture probes needs to be experimentally determined for each application (*see* **Note 8**).

For direct analysis of nucleic acids of high genetic complexity, the sample should be fragmented by sonication, restriction digestion, or chemical cleavage before being hybridized with the array of capture probes.

For direct transcriptional profiling, mRNA is extracted from the biological sample, optionally converted to cDNA, heat denatured, and mixed with the desired set of gene-specific labeled stacking probes (typically 15–30 bases in length), then hybridized with the appropriate set of arrayed capture probes (typically 7–10 bases long) specifically designed to hybridize to the target strand in tandem with the labeled probes.

4. Notes

1. The results reported by the Mirzabekov laboratory indicated that contiguous stacking hybridization may be a viable strategy to resolve sequence ambiguities in sequencing by hybridization *(1,3,4)* and to identify specific point mutations *(2)*. The contiguous stacking hybridization strategies described by the Mirzabekov group require multiple rounds of hybridization, whereas the tandem hybridization strategy described herein is designed to detect known sequence variations, in a single hybridization reaction on the oligonucleotide array. Although the tandem hybridization approach described herein has a more limited range of applications than Mirzabekov's contiguous stacking hybridization, it is simpler, yet applicable to many important DNA diagnostic tests, in which the relevant alleles are known from previous research.

2. The 3'-C3-amine group is introduced into oligonucleotides using a special controlled pore glass (CPG) synthesis support designed to introduce a 3'-amino group onto synthetic oligonucleotides: the 3'-Amino-Modifer C3 CPG available from Glen Research or the 3'-Amine-ON CPG from Clonetech. On completion of the solid-phase synthesis, the oligonucleotide-bearing CPG is incubated in ammonia to cleave the succinate linkage between the oligonucleotide and glass surface. Then, on full deprotection in concentrated ammonia, the free 3'-terminal amine is generated (*see* **Fig. 2**), and thus, the 3' modification is actually an aminopropanol function. It is critical to specify the C3 3' amine modification when ordering the oligonucleotides. Direct attachment to underivatized glass does not occur with the C7 3'-amine modification, nor with the C6 5'-amine modification.

3. The use of synthetic oligonucleotide targets is recommended as a positive hybridization control and for optimizing hybridization conditions.

4. Using this simple attachment method, the entire process of cleaning the slides, attaching oligonucleotide probes, prehybridization, and hybridization can be easily carried out in a single workday. Quantitation of 3'-aminopropanol-derivatized oligonucleotide binding to glass, using a phosphorimager, indicates that oligonucleotides are spaced approx 50–100Å apart on the glass surface, equivalent to 10^{10}–10^{11} molecules/mm^2. Although the structure of the linkage is not known with certainty, it is proposed to be an ester linkage between the silanol group on the glass and the hydroxyl group on the terminal aminopropanol function of the oligonucleotide, as shown in **Fig. 3**. This ester linkage, rather than the amide linkage, is suggested by the observations that the linkage is stable in hot water and in mild acid but labile in mild base, and not formed with 5'-alkylamino-derivatized oligonucleotides (indicating the requirement of the hydroxyl function). The amino function, though apparently not the point of attachment, nevertheless plays some role, because the attachment reaction is blocked by acetylation of the terminal amine *(16,17)*.

5. For a 200-bp PCR product 0.2–0.5 pmol corresponds to approx 27–67 ng.

6. The stacking probe preannealed with the target nucleic acid must be long enough to enable base stacking forces to propagate through the stacking probe into the capture probe, stabilizing the capture of target sequence to the short capture

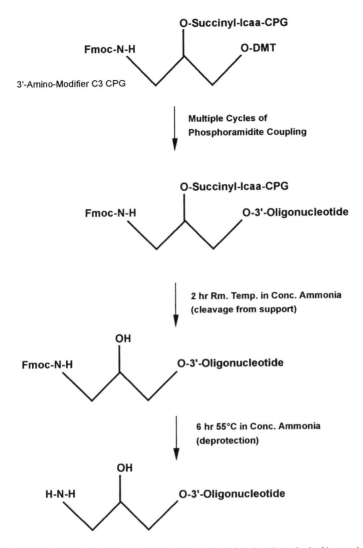

Fig. 2. Chemical synthesis of oligonucleotides derivatized at their 3' termini with the aminopropanol (propanolamine) function. The standard phosphoramidite procedure is followed, using as solid-phase support the 3'-Amino-Modifier C3 CPG (Glen Research). The structure of this support material is shown at the top. The aminopropanol group is covalently attached to the glass via long chain alkylamine-succinate linkage at carbon two, the amino group is proteced by the Fmoc group, and the hydroxyl group is protected by the dimethoxytrityl (DMT) group. Multiple cycles of DMT-phosphoamidite coupling are carried out to build up the desired 5'-DMT-protected oligonucleotide sequence. The 5'-DMT protecting group is removed by acetic acid treatment (as done in the first step of each synthesis cycle). Then the succinate linkage is cleaved by a 2-h incubation at room temperature in concentrated ammonia, releasing the 3'-amine-protected oligonucleotide from the support. The Fmoc protecting group (together with exocyclic protecting groups on the bases) are cleaved during incubation in concentrated ammonia at 55°C for 6 h, yeilding the 3'-aminoproanol-derivatized oligonucleotide.

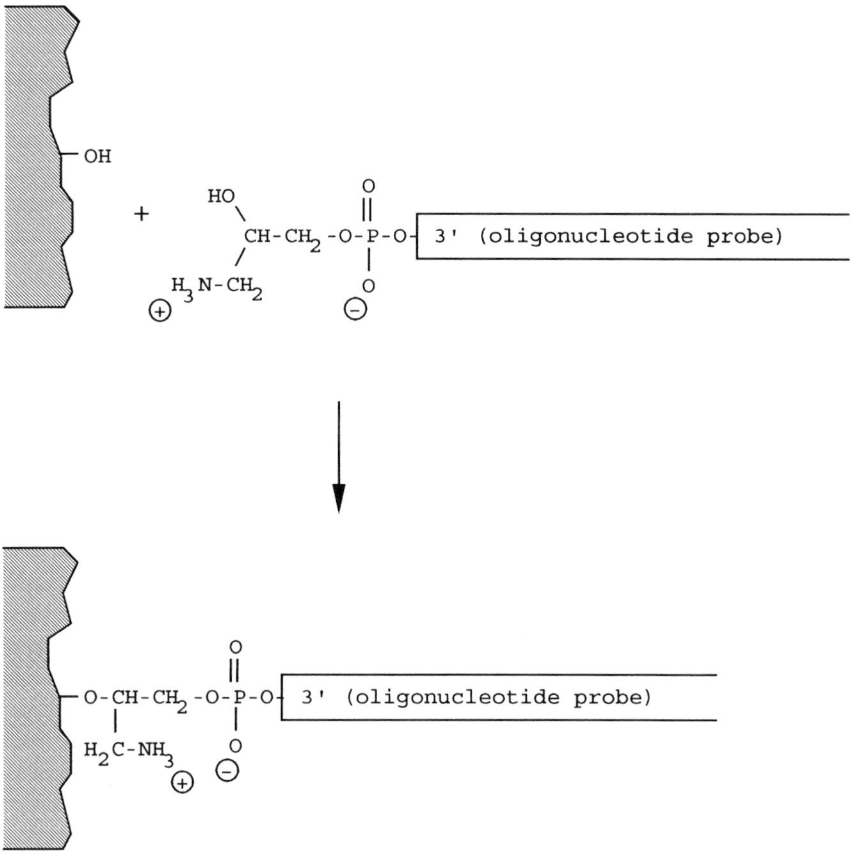

Fig. 3. Proposed structure of the linkage formed between 3'-aminopropanol-derivatized oligonucleotide and underivatized glass. The aminopropanol function at the 3' terminus of the oligonucleotide is shown at the top, together with the –OH (silanol) function on the surface of the glass. The proposed ester linkage formed between the silanol group on the glass and C2-OH of the oligonucleotide's 3'-aminopropanol function is shown at the bottom.

probe. Stacking probes 20–30 bases in length are recommended, although we have not characterized the stabilization as a function of stacking probe length. The appropriate length of labeled stacking probe also depends on the genetic complexity of the nucleic acid being analyzed. The stacking probe must be long enough to bind to a single unique site within the target nucleic acid. The formula $n = L/4^p$ can be used to predict the average number of times, n, a given oligonucleotide probe of length, p, will bind to a target sequence of genetic complexity (total length of unique nonrepeated sequence), L. Using this formula, it can be predicted that an array of 8mer probes would not serve to analyze a total mRNA mixture extracted from bacterial cells using the traditional hybridization approach wherein each target sequence is specified by a single surface-tethered probe: Each 8-mer probe would be predicted to occur, on average, about eight times in a nucleic acid mixture of genetic complexity, 5×10^5, representative of bacterial RNA. In this case every capture probe would hybridize with multiple sequences within the RNA sample, masking the desired gene-specific transcriptional profile. In the tandem hybridization strategy, however, the effective probe length is greater than that of the capture probe alone, since hybridization depends on contiguous stacking of capture probe with the tandemly hybridizing labeled stacking probe. In the experience of coauthor Beattie's laboratory (Fleming and Beattie, unpublished data), 8-mer capture probes can indeed be used, in combination with tandemly hybridizing labeled 20-mer stacking probes, to observe gene induction with total RNA extracted from bacteria. For analysis of bulk RNA from mammalian cells, capture probes of 9-mer or 10-mer length are recommended, and the hybridization temperature should be increased accordingly (*see* **Note 8**).

7. TMAC is commonly used to minimize the effect of base composition on duplex stability. In the tandem hybridization strategy, however, the stabilization provided by base stacking interactions propagated from the long stacking probe into the short capture probe may itself serve to minimize the influence of base composition. In coauthor Beattie's laboratory, TMAC is routinely replaced by 6X SSC in the hybridization buffer used in the tandem hybridization method.

8. The optimum hybridization temperature depends on the length and base composition of the capture probe, and may be additionally influenced by sequence-dependent base stacking interactions around the junction of the capture and stacking probes. As a general rule, with long stacking probes (20- to 30-mer), the recommended hybridization temperature is 15°C for 7-mer capture probes containing four [G + C] and 25°C for 9-mer capture probes containing five [G + C]; however, some experimentation may be required to define the optimum hybridization temperature for a given set of capture probes. Furthermore, in the design of pairs of stacking and capture probes for a given application, it is desirable to adjust the position of probes along the target sequence and/or the length of capture probe, to achieve a narrow range of hybrid stability (or predicted T_m) across the entire array of capture probes. In practice, it is sometimes necessary to test several probes of various length and sequence for each target sequence of interest, to define an optimal set of probes for a given oligonucleotide microarray application.

References

1. Parinov, S., Barsky, V., Yershov, G., Kirillov, E., Timofeev, E., Belgovskiy, A., and Mirzabekov, A. (1996) DNA sequencing by hybridization to microchip octa- and decanucleotides extended by stacked pentanucleotide. *Nucleic Acid Res.* **24,** 2998–3004.

2. Yershov, G., Barsky, V., Belgovskiy, A., Kirillov, E., Ivanov, I., Parinov, S., Guschin, D., Drobishev., Dubiley, S., and Mirzabekov, A. (1996) DNA analysis and diagnostics on oligonucleotide microchips. *Proc. Natl. Acad. Sci. USA* **93**, 4913–4918.

3. Khrapko, K. R., Lysov, Y. P., Khorlin, A. A., Schick, V. V., Florentinev, V. L., and Mirzabekov, A. D. (1989) An oligonucleotide hybridization approach to DNA sequencing. *FEBS Lett.* **256**, 118–122.

4. Khrapko, K. R., Lysov, Y. P., Khorlin, A. A., Ivanov, I. B., Yershov, G. M., Vasilenko, S. K., Florentiev, V. L., and Mirzabekov, A. D. (1991) A method for DNA sequencing by hybridization with oligonucleotide matrix. *DNA Sequence* **1**, 375–388.

5. Kieleczawa, J., Dunn, J. J., and Studier, F. W. (1992) DNA sequencing by primer walking with strings of contiguous hexamers. *Science* **258**, 1787–1791.

6. Kotler, L. E., Zevin-Sonkin, D., Bobolev, I. A., Beskin, A. D., and Ulanovsky, L. E. (1993) DNA sequencing: modular primers assembled from a library of hexamers or pentamers. *Proc. Natl. Acad. Sci. USA* **90**, 4241–4245.

7. Kaczorowski, T. and Szybalski, W. (1994) Assembly of 18-nucleotide primers by ligation of three hexamers: sequencing of large genomes by primer walking. *Anal. Biochem.* **221**, 127–135.

8. Kaczorowski, T. and Szybalski, W. (1996) Co-operativity of hexamer ligation. *Gene* **179**, 189–193.

9. Lodhi, M. A. and McCombie, W. R. (1996) High-quality automated DNA sequencing primed with hexamer strings. *Genome Res.* **6**, 10–18.

10. Johnson, A. F., Lodhi, M. A., and McCombie, W. R. (1996) Fluorescence-based sequencing of double-stranded DNA by hexamer string priming. *Anal. Biochem.* **241**, 228–237.

11. Maldonado-Rodriguez, R., Espinosa-Lara, M., Calixto-Suárez, A., Beattie, W. G., and Beattie, K. L. (1999) Hybridization of glass-tethered oligonucleotide probes to target strands preannealed with labeled auxiliary oligonucleotides. *Mol. Biotechnol.* **11**, 1–12.

12. Maldonado-Rodriguez, R., Espinosa-Lara, M., Loyola-Abitia, P., Beattie, W. G., and Beattie, K. L. (1999) Mutation detection by stacking hybridization on genosensor arrays. *Mol. Biotechnol.* **11**, 13–25.

13. Beattie, K. L., Beattie, W. G., Meng, L., Turner, S. L., Coral-Vazquez, R., Smith, D. D., McIntyre, P. M., and Dao, D. D. (1995) Advances in genosensor research. *Clin. Chemistry* **41**, 700–706.

14. Beattie, W. G., Meng, L., Turner, S. L., Varma, R. S., Dao, D. D., and Beattie, K. L. (1995) Hybridization of DNA targets to glass-tethered oligonucleotide probes. *Mol. Biotechnol.* **4**, 213–225.

15. Beattie, K. L., Zhang, B., Tovar-Rojo, F., and Beattie, W. G. (1996) Genosensor-based oligonucleotide fingerprinting, in *Pharmacogenetics: Bridging the Gap Between Basic Science and Clinical Application* (Schlegel, J., ed.), IBC Biomedical Library, Southborough, MA, pp. 5.1.5–5.1.25.

16. Doktycz, M. J. and Beattie, K. L. (1997) Genosensors and model hybridization studies, in *Automated Technologies for Genome Characterization* (Beugelsdijk, A. J., ed.), Wiley, New York, pp. 205–225.

17. Beattie, K. L. (1997) Analytical microsystems: Emerging technologies for environmental biomonitoring, in *Biotechnology in the Sustainable Environment* (Sayler, G., ed.), Plenum Publishing, New York, pp. 249–260.

18. Beattie, K. L. (1997) Genomic fingerprinting using oligonucleotide arrays, in *DNA Markers: Protocols, Applications and Overviews* (Caetano-Anolles, G. and Gresshoff, P., eds.), Wiley, New York, pp. 213–224.

19. Matteucci, M. D. and Caruthers, M. H. (1981) Synthesis of deoxyoligonucleotides on a polymer support. *J. Am. Chem. Soc.* **103,** 3185–3191.

11

DNA Sequencing by Hybridization with Arrays of Samples or Probes

Radoje Drmanac, Snezana Drmanac, Joerg Baier, Gloria Chui,
Dan Coleman, Robert Diaz, Darryl Gietzen, Aaron Hou, Hui Jin,
Tatjana Ukrainczyk, and Chongjun Xu

1. Introduction

This chapter focuses on sequencing by hybridization (SBH), an advanced DNA sequencing technique first proposed in 1987 *(1)*. SBH procedures determine DNA sequence information by screening DNA oligomers (typically 7- to 11-mers) for their ability to hybridize with target DNA. The set of overlapping oligomers that matches the target DNA is then used to assemble its sequence. The theory, practice, and history of SBH are reviewed in **refs.** *2* and *3*.

There are three established implementations of SBH methods, each of which relies on different experimental procedures and strategies. The principle common to these three formats of SBH is the use of overlapping probe sequences to compile a target sequence. In Format 1 and its variant Format 1A SBH *(4)*, target DNA molecules (typically 1–2 kb in length) are attached to a solid support such as a GeneScreen membrane. A complete or selected set of labeled oligomer probes is exposed to the target DNAs and then membranes are washed to remove unbound probe. Those probes that hybridize to the target DNA are then read in an appropriate device such as a phosphorimager or fluorescence reader. In Format 3 SBH *(3)*, both bound and free labeled probes are used. When one bound and one labeled (free) probe hybridize to the target DNA at precisely adjacent positions, they are covalently linked in the presence of DNA ligase to form a longer, support-bound, labeled nucleic acid. A detector device appropriate for the labeled probes is then used to read their labels.

In all three SBH formats, the raw data can be processed by special software programs that locate positions within the array, normalize and rank signal

From: *Methods in Molecular Biology, vol. 170: DNA Arrays: Methods and Protocols*
Edited by: J. B. Rampal © Humana Press Inc., Totowa, NJ

intensities, and then determine DNA sequences by compiling the sequences of the subset of overlapping probes that hybridize each target DNA. Large probe sets may be used to completely sequence target DNAs, whereas smaller sets of probes may be used to fingerprint, map, or partially sequence target DNAs *(5)*.

This chapter focuses on the basic materials and methods used in Formats 1 and 3 SBH. For more detailed information on these methods, *see* **refs. *3*, *4*,** and ***6*.**

2. Materials

2.1. Format 1 SBH

2.1.1. Polymerase Chain Reaction Procedures for DNA Samples

1. Peltier Thermal Cycler Model 100 or 200 (MJ Research, Waltham, MA).
2. MJ Research polymerase chain reaction (PCR) tubes.
3. AmpliTaq DNA polymerase (5 U/µL) (PE Biosystems, Foster City, CA) is diluted to 0.025 U/µL for PCR.
4. Plaque-forming unit PCR buffer.

2.1.2. Binding of DNA Samples to Solid Support

1. GeneScreen membranes (New England Nuclear, Boston, MA).
2. Gridding robot with three-axis gantry (in-house designed).
3. 0.5 *M* NaOH.
4. Ultraviolet (UV)-crosslinker (Stratagene, La Jolla, CA).

2.1.3. Probes

Probes are purchased from Biosource (Palo Alto, CA) or GenSet (San Diego, CA).

2.1.4. Probe Labeling, Hybridization, and Washing

1. γ^{33}P-ATP (0.125 µL) (10 mCi/mL) (Amersham, Piscataway, NJ).
2. T4 polynucleotide kinase (0.05 mL) (30 U/µL) (Amersham, Piscataway, NJ).
3. 384-well plates (NUNC or CoStar).
4. 10X kinase buffer (2.0 µL).
5. Hybridization buffer: 0.2 *M* sodium phosphate, 6% lauryl sarcosine.
6. Probes (2.5 µL @ 2 ng/µL) are pipetted using a Biomek Automated Laboratory Workstation (Beckman Coulter, Fullerton, CA) with a pipetting tool.
7. Wash buffer: 4X saline sodium citrate.
8. Phosphor screens (Molecular Dynamics, Sunnyvale, CA).
9. PhosphorImager® scanner (Molecular Dynamics).

2.1.5. Scoring and Analysis of Hybridized Arrays

Array images are analyzed by proprietary image analysis programs developed at Hyseq (*see* **Note 1**).

2.2. Format 3 SBH

2.2.1. Construction and Binding of 5-mer Probes to Solid Supports

1. Pentamer probes are obtained from PE Biosystems (Foster City, CA) or are obtained from Biosource or GenSet.
2. Probes are stored at 200 µ*M* in H_2O in 96-well 2-mL deep-well plates (VWR 4000-012).
3. Glass microscope slides (Erie Scientific Soda Lime Glass, Portsmouth, NH) are derivitized at Hyseq.
4. Eight-channel pipetting tool and an IAI robot (Automation Controls Group, Campbell, CA).

2.2.2. Construction of Labeled Probes

1. TAMRA fluorescent-labeled probes are obtained from PE Biosystems.
2. Labeled probes are diluted to 100 µ*M* in H_2O.

2.2.3. PCR Procedures for DNA Samples

1. Genomic DNA.
2. MJ Research PTC-200 Peltier Thermal Cycler.
3. 10X PCR buffer: 10 m*M* Tris-HCl, pH 8.3, 50 m*M* KCl.
4. *Taq* polymerase (5 U/µL) is added to the 2.5 U/100 µL reaction mixture (PE Biosystems).
5. Lambda exonuclease (λ exo) (Gibco-BRL, Rockville, MD).
6. DNAse endonuclease (Gibco-BRL).

2.2.4. Hybridization and Ligation of Target DNA and Free Probes

1. Ligation mixture: 2–5 U/µL of T4 DNA ligase (New England Biolab, Beverly, MA), 10% PEG 8000, 5 m*M* spermidine (Sigma, St. Louis, MO), 2–10 n*M* target DNA, 5–40 n*M* labeled probe (each), 1X ligase buffer.
2. SSPE wash buffer (sodium chloride, sodium phosphate, EDTA).

2.2.5. Scoring and Analysis of Ligated 10-mer Probes

1. Axon GenePix 4000A Fluorescence Reader (Axon, Foster City, CA) or Scan Array® 3000 (GSI Lumonics, Watertown, MA).
2. Image data are analyzed and the sequence is compiled using proprietary software programs developed at Hyseq (*see* **Note 1**).

3. Methods

3.1. Format 1 SBH

3.1.1. PCR Procedures for DNA Samples

For each pair of primers, at least 35 PCR cycles are performed under optimized conditions using small amounts of bacterial cultures containing the appropriate vector/cDNA inserts (*see* **Note 2**).

3.1.2. Binding of DNA Samples to Solid Support

1. Dispense PCR samples from microtiter plates onto nylon membranes using a robot equipped with a pin tool. The pin tool transfers approx 20 nL to each spot, for a total DNA mass of 0.5 ng/spot.
2. Soak membranes for at least 10 s in 0.5 M NaOH solution to denature DNA samples.
3. After spotting, allow membranes to dry for 3 h at room temperature and subsequently UV-irradiate at 1200 J to attach or fix the DNA to the membrane (*see* **Note 3**).

3.1.3. Probes

Probes used in SBH experiments are pools of oligomers, each oligomer having a specific core sequence, surrounded by two to four bases that vary among the oligomers of a given pool, e.g., N-B7-NN (*see* **Note 4**). Deprotected, desalted probes are arrayed in 96-tube racks.

3.1.4. Probe Labeling, Hybridization, and Washing

1. Label 10 ng of each probe with γ^{33}P-ATP in 384-well plates under conditions designed to phosphorylate at least 50% of the DNA molecules. An entire membrane may be exposed to a single probe, or individual DNA microarrays may be exposed to different probes by using a specially designed grid to create physically separate hybridization chambers. In the latter case, add hybridization buffer and an aliquot of probe (0.8 pmol/mL) to each chamber using a robotic pipetting device.
2. Hybridize and wash probes under low temperature (0–5°C) for approx 30 min.
3. After washing, blot the membranes on filter paper and place in cassettes containing phosphor screens for 1 h at 4°C before reading with a phosphorimager.

3.1.5. Scoring and Analysis of Hybridized Arrays

The following software programs have been developed by Hyseq programmers to analyze Format 1A image data:

1. Image analysis: locates the physical position of probes hybridized to spotted samples fixed in an array and assigns a score to each probe.
2. Mass normalization: compensates for mass differences among different sample spots in an array.
3. Sequence analysis: calculates a positive probe frequency for each base position in the sample relative to the reference sequence that indicates whether the base at that position is a wild-type, heterozygous, or homozygous base and determines the most probable base for each sample and position.

3.2. Format 3 SBH

3.2.1. Synthesis and Binding of 5-mer Probes to Solid Support

Probes are transferred to a glass support using a tool and a laboratory robot. They are spotted in predefined arrays and are fixed to the slide by covalent attachment of modified oligonucleotides to derivatived glass surfaces.

In Format 3 SBH, multiple unit arrays of probes are prepared on one glass substrate, similar to the arrays of unit sample arrays used in Format 1 SBH. These replica arrays are used to hybridize different labeled probes with the same sample DNA (*see* **Note 5**).

3.2.2. Construction of Labeled Probes

Commercially obtained fluorescent tags are attached at the 3' end of the 5-mer probe.

3.2.3. PCR Procedures for DNA Samples

Target DNA is amplified from genomic DNA using primers designed by the Primer 3 Input Web site program and standard PCR procedures, as follows:

1. Add 6 ng of genomic DNA and 1–3 ng/µL of each primer to 100 µL of PCR buffer.
2. PCR conditions: 96°C for 3 min; 42 cycles at 96°C for 1 min, 57–64°C for 45 s to 1 min, 72°C for 1.5 min; then hold at 72°C for 5 min. Specific PCR conditions were adapted to suit particular DNA samples.
3. Pool PCR products from the initial PCR reactions and mix 1 µL of a 20- to 50-fold dilution of the pooled mixture with 1 or 3 ng/µL of each primer in a second 100-µL PCR reaction.
4. Verify PCR products by agarose gel electrophoresis using marker DNA ladders.
5. Single-stranded DNA may be prepared by degradation of the strand containing the phosphorylated primer or by asymmetric PCR. Digest PCR-amplified double-stranded DNA with λ exo to remove the phosphorylated strand. Exonuclease is inactivated by heating the mixture to 96°C for 5 min. Use agarose gel electrophoresis to confirm the formation of single stranded DNA.
6. To increase availability of the target for hybridization, fragment the DNA to short pieces (on average 20–40 bases) by endonuclease cleavage using an optimized DNaseI concentration for each batch of enzyme at 37°C for 30 min.

3.2.4. Hybridization and Ligation of Target DNA and Labeled Probes

1. Add target DNA and a labeled 5-mer (or longer) probe or a pool of probes (*see* **Note 6**) to the glass slide hybridization chamber containing the complete set of 1024 attached 5-mer probes, plus assorted positive and negative controls. First, preheat the target DNA to 98°C for 10 min, and then add to the hybridization mix

in the absence of DNA ligase. Heat the mixture to 98°C for an additional 10 min, cool suddenly, then add ligase and pipet mixture onto array surface.

2. Incubate the reaction at 4°C or a higher temperature that is suitable for the particular probe, target, and enzyme concentrations used. Incubate the mixture until most of the target DNA has been used as template (*see* **Note 7**).
3. Wash the slides in SSPE buffer.
4. Scan the slides at 20-μ pixel resolution.

3.2.5. Scoring and Analysis of Ligated 10-mer Probes

Image analysis, including score normalization of positive probe frequency, is done using special software programs developed at Hyseq. Format 3 SBH methods may also be adapted for mutation or polymorphism discovery (*see* **Note 8**).

4. Notes

1. Formats 1 and 3 image analysis programs were developed at Hyseq, but other commercially available image analysis software may be used.
2. Specific PCR conditions are selected to suit individual DNA samples, but typically involve 40 cycles at 94°C for 15 s, 55°C for 15 s, and 72°C for 90 s using 400 ng of each primer and 2 U of *Taq* I polymerase/100 μL.
3. Many DNA samples may be spotted in a single array; Hyseq routinely processes Format 1 arrays containing more than 51,840 clones of interest using a robot equipped with a 384-pin tool. Additional controls and blanks bring the total to more than 55,000 spots per membrane.
4. Most 7-mer probe sequences consists of pools of 10- to 11-mer probes with a specific 7-mer core and a few variable bases at the ends, e.g., N-B7-NN, in which N may represent any of the four DNA bases and B7 indicates a specific seven-base sequence. Hence, a 7-mer probe sequence with three variable bases would consist of a pool of 64 possible 10-mers, each with an identical 7-mer core sequence. The variable bases elongate the probe, increasing the thermodynamic stability of the DNA-probe hybrid.
5. In preparing probe arrays, probe DNA is attached to the solid substrate. A saturating quantity of probe is used to reduce the quantity of target DNA and/or labeled probe that is required to produce a strong signal. Preparing arrays from high-quality premade probes has many advantages compared to *in situ* synthesis of arrays of probes, which inevitably results in many defective probes.
6. To efficiently score all possible 10-mers, the labeled probes are added to the unit arrays in pools containing 16 or more probes tagged with the same dye. Larger pools using two or more colors may also be used to reduce the number of hybridization steps. The pool size selected is inversely proportional to the target length. Because of the low frequency of positive 10-mers in targets of length 200–2000 bases, pools with hundreds of probes may be used.
7. Labeled probes that hybridize to the template DNA at positions exactly adjacent to attached probes are potential targets for DNA ligase, which covalently links

the two probes to form a labeled 10-mer attached to the support. These 10-mers give a strong signal in a fluorescence array scanner, confirming the presence of a complementary 10-mer sequence within the target DNA.

8. In addition to full-length sequencing, the Format 3 process may be used for mutation or polymorphism detection. The probe selection procedure is similar to that used in Format 1 SBH. A subset of labeled probes corresponding to the reference sequence of a target DNA is selected and arranged in four or more pools, and each pool is hybridized to a test sample on a complete array of attached 5-mer probes. Pools are assembled to avoid ligation of two different labeled probes from one pool with a single attached probe. This allows researchers to assign correctly a positive signal to the appropriate 10-mer. A sudden drop in positive 10-mer probe frequency at a given base indicates that a mutation or polymorphism at that region is present in that DNA sample.

References

1. Drmanac, R. and Crkvenjakov, R. (1987) Method of sequencing of genomes by hybridization with oligonucleotide probes. Yugoslav Patent Application P570/87. US Patent No. 5,202,231 (1993).

2. Drmanac, R. and Drmanac, S. (2000) Sequencing by hybridization arrays, in *DNA Arrays: Methods and Protocols: Methods in Molecular Biology* (Rampal, J. B., ed.), Humana, Totowa, NJ, pp. 39–51.

3. Drmanac, R., Drmanac, S., and Little, D. (2000) Sequencing and fingerprinting DNA by hybridization with oligonucleotide probes, in *Encyclopedia of Analytical Chemistry*, John Wiley & Sons Ltd., pp. 5232–5257.

4. Drmanac, S., Kita, D., Labat, I., Hauser, B., Burczak, J., and Drmanac, R. (1998) Accurate sequencing by hybridization for DNA diagnostics and individual genomics. *Nat. Biotechnol.* **16,** 54–58.

5. Drmanac, R. and Drmanac, S. (1999) cDNA screening by array hybridization. *Meth. Enzymol.* **303,** 165–178.

6. Drmanac, R., Drmanac, S., Strezoska, Z., Paunesku, T., Labat, I., Zeremski, M., Snoddy, V., Funkhouser, W. K., Koop, B., Hood, L., and Crkvenjakov, R. (1993) DNA Sequence determination by hybridization: A strategy for efficient large-scale sequencing. *Science* **260,** 1649–1652.

12

Using Oligonucleotide Scanning Arrays to Find Effective Antisense Reagents

Muhammad Sohail and Edwin M. Southern

1. Introduction

Antisense oligonucleotides (AONs) are synthetic deoxyribonucleic acids, typically between 15–25 nucleotides, that can bind to complementary sites in a mRNA and inhibit translation. AONs show promise as therapeutics to many diseases caused by abnormal or unwanted expression of a gene, such as cancers. The first antisense oligonucleotide-based drug (Vitravene™: ISIS Pharmaceuticals) was introduced in 1998 to treat cytomegalovirus (CMV) retinitis in AIDS patients: several others are in advanced stages of clinical trials for viral diseases and cancer. Antisense oligonucleotides are also a useful tool in the functional analysis of genes.

The selection of an effective target sequence is a considerable obstacle to wider application of the antisense technology. The ability of an AON to bind to the target mRNA determines its efficacy as an antisense reagent. Invariably, only a few of the oligonucleotides complementary to a target mRNA is found to be effective antisense reagents. Oligonucleotides targeted to regions separated only by a few bases can have markedly different antisense effects *(1–3)*.

There is compelling evidence that duplex formation is constrained by the folded structure of RNAs and is not determined primarily by base sequence or composition *(2,3)*. Several computational methods are available for prediction of secondary structure of RNAs calculated from thermodynamic properties and nearest-neighbor interaction *(4,5)*, but they are generally believed to be unreliable in predicting folding of large RNAs and thus have a limited utility in the selection of AONs *(6,7)*. The rules that govern heteroduplex formation are also poorly understood. In a recent study, using arrays of antisense oligonucleotides,

From: *Methods in Molecular Biology, vol. 170: DNA Arrays: Methods and Protocols*
Edited by: J. B. Rampal © Humana Press Inc., Totowa, NJ

targeted to structurally well characterized tRNA^{Phe}, Mir, and Southern *(3)* showed that it is quite difficult to derive rules to predict effective target sites on RNAs. Therefore, even with better understanding of secondary structure, more insight into the mechanisms of heteroduplex formation will be needed.

Inability to predict effective antisense target site indicates a need to develop empirical screens and several methods have been developed (for review, *see* **refs.** *8* and *9*). We use AON scanning arrays to find oligonucleotides that have high binding affinity to target nucleic acids. Scanning arrays are a simple tool that allow combinatorial synthesis of a large number of oligonucleotides on a solid platform (typically glass or polypropylene: *see* **Note 1**) in a spatially addressable fashion, and parallel measurement of the binding of all oligonucleotides complementary to the target mRNA *(1)*.

The scanning arrays comprise sets of oligonucleotides of various lengths. A series of oligonucleotides, complementary to the target mRNA, is made by sequential coupling of nucleotides to a solid surface. The DNA synthesis reagents are applied to a confined area on the surface of the solid support using a mask (*see* below). The mask is shifted along the surface after each round of coupling, resulting in a series of oligonucleotides each complementary to a region of the target sequence.

Using a diamond-shaped or a circular reaction mask (*see* **Fig. 1**), it is possible to create arrays comprising sets of oligonucleotides of all lengths from monomers up to a maximum in a single series of couplings. The maximum length of oligonucleotides synthesised depends upon the ratio of the diagonal (for a diamond-shaped mask) or diameter (for a circular mask) of the mask to the displacement at each coupling step. For example, a diamond-shaped mask of 40 mm diagonal will produce 10-mers, 16-mers, or 20-mers using step sizes of 4, 2.5, or 2 mm, respectively. A diamond-shaped template creates a series of small diamond-shaped cells. The longest oligonucleotides are found along the center line and the monomers are located at the edge (*see* **Fig. 1**). The circular template creates cells that differ in shape: along the center line, they are lenticular, but off this line, they form a four-cornered spearhead that diminishes in size toward the edge. The arrays as synthesized are symmetrical above and below the center line of the template and each oligonucleotide is represented twice allowing for duplicate hybridization measurements.

For each length of oligonucleotides s, there are $N - s + 1$ s-mers covering a total length of N bases. For example, if a 150 nt long sequence is covered in a 150-step synthesis, there will be 150 monomers and 131 20-mers. The last 20 positions in the sequence will be represented by shorter oligonucleotides only; in this case, from 19-mer to monomer. Therefore, for making 200 20-mers, an additional 19 nt synthesis steps need to be added at the end, i.e., total coupling steps $= N + s - 1$.

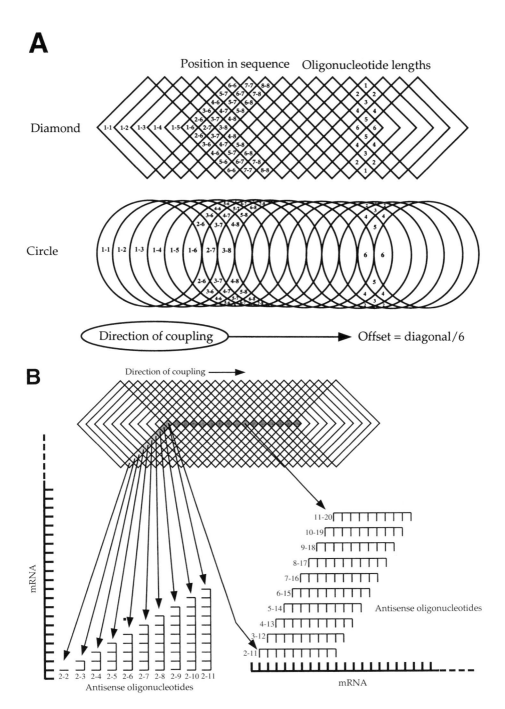

Fig. 1. Layout of scanning arrays. (**A**) Two template shapes are illustrated; a diagonal shape created by turning a square through 45° and a circle. (**B**) An illustration showing lengths and position of oligonucleotides on an array made using a diamond-shaped mask.

This chapter describes the methods of fabricating reaction masks and scanning arrays, preparation of radiolabeled targets, hybridization to the arrays, and computer-aided analysis of the hybridization data (*see* **Fig. 2**).

2. Materials

2.1. Derivatization of Glass

1. Glass cylinder and apparatus shown in **Fig. 3.**
2. Glass sheets of required dimension (3 mm thick: Pilkington, UK).
3. 3-Glycidoxypropyl trimethyoxysilane (98% [v/v]: Aldrich).
4. Di-isopropylethylamine (99.5% v/v: Aldrich).
5. Xylene (AnalaR: Merck).
6. Hexaethylene glycol (97% [v/v]: Aldrich).
7. Sulfuric acid (AnalaR: Merck).
8. Ethanol (AnalaR: Merck).
9. Ether (AnalaR: Merck).
10. Water bath at 80°C.

2.2. Making Reaction Masks

1. Stainless steel or aluminium square metal piece or polytetrafluoroehtylene (PTFE) (Teflon). Dimensions of the workpiece may vary according to the size of the mask.
2. A center lathe or a horizontal milling machine.
3. A drilling machine.
4. Abrasive paper from approx P600 to P1200 (3M Inc., USA) and polishing-grade crocus paper (J. G. Naylor & Co. Ltd., Woodley, Stockport, Manchester, England).

2.3. Making Scanning Arrays

1. Solid support (derivatized glass or aminated polypropylene [Beckman Coulter]).
2. DNA synthesizer (ABI).
3. A reaction mask of desired shape and size and assembly frame (*see* **Fig. 4** for the assembly).
4. DNA synthesis reagents: standard dA, dG, dC, and T phosphoramidites; oxidizing agent; acetonitrile; activator solution; deblock solution (all from Cruachem).
5. Reverse phosphoramidites bought (Glen Research).

2.4. Deprotection of Arrays

1. Assembly for constructing deprotection bomb as shown in **Fig. 5**. The assembly consists of a high-density polyethylene (HDPE) chamber, 4 mm thick silicon rubber gasket and a stainless steel plate of the dimensions of the HDPE chamber, and stainless steel M8 nuts and bolts.
2. 30% Ammonia solution (AnalaR: Merck).
3. Water bath at 55°C.

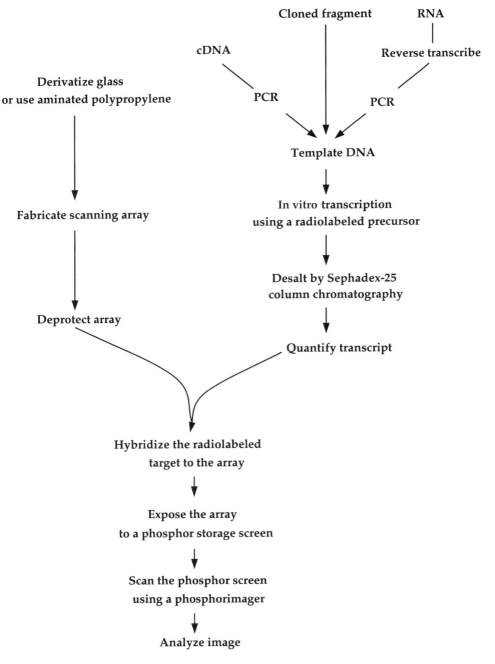

Fig. 2. Flowchart.

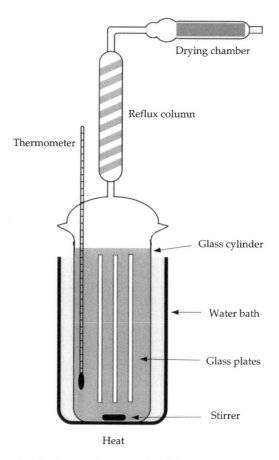

Fig. 3. Assembly for derivatizing glass plates.

Fig. 4. *(opposite page)* (**A**) Apparatus for applying reagents to the surface of solid support. Upper: A diamond-shaped metal mask. The mask consists of a 50×50 mm metal block with a diamond-shaped sealing edge, 0.5 mm height $\times$ 50 mm diagonal (drawings not to scale). The mask is fixed to a frame, which carries the lead screw shown in the lower panel. Lower: The mask is mounted on a frame made from 50 mm $\times$ 25 mm angle aluminium, fixed to the front of a DNA synthesizer. The lead screw, 1.0-mm pitch, is fitted with a pusher nut that drives the plate across the front of the mask. The driving wheel is marked in half-turns to enable the plate to be incremented forward in half millimeter steps. The pressure clamp is a modified engineer's G-clamp fixed to the back of the frame with a polyethylene cushion mounted on its pressure pad. (**B**) One coupling cycle comprises clamping the plate up to the mask, starting the DNA synthesiser to go through the preprogrammed cycle to couple the appropriate nucleotide, slackening the clamp, and moving the plate one increment by driving the lead screw the desired number of whole or half-turns.

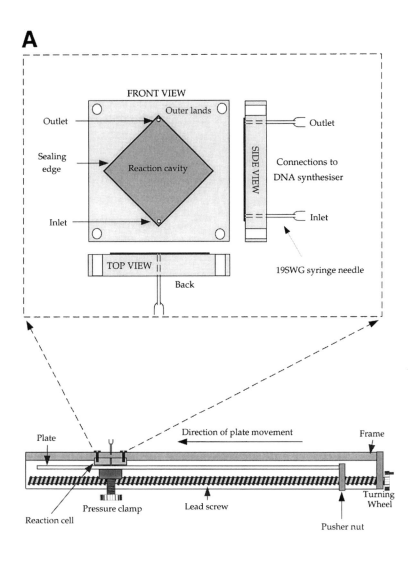

A

FRONT VIEW

Outer lands

Outlet

Sealing
edge

Reaction cavity

Inlet

SIDE VIEW

Outlet

Connections to
DNA synthesiser

Inlet

19SWG syringe needle

TOP VIEW

Back

Plate

Direction of plate movement

Frame

Pressure clamp

Lead screw

Turning
Wheel

Reaction cell

Pusher nut

B

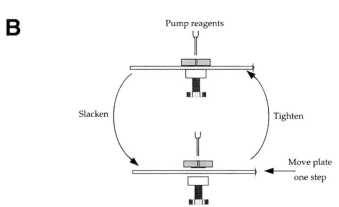

Pump reagents

Slacken

Tighten

Move plate
one step

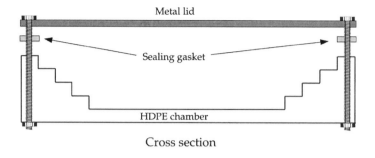

Cross section

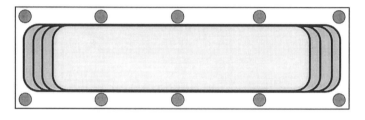

Top view

Fig. 5. Illustration of the assembly used for deprotecting arrays.

2.5. In Vitro Transcription

1. Template DNA (at approx 1 mg/mL).
2. T7 or SP6 RNA polymerase, transcription buffer (5X transcription buffer is 200 mM Tris-HCl, pH 7.9, 30 mM MgCl$_2$, 10 mM spermidine, 50 mM NaCl), recombinant RNAsin®, 100 mM dithiothreitol (DTT) and nuclease-free distilled water (Promega).
3. [α-^{32}P] UTP (3000 Ci/mmol) or [α-^{33}P] UTP (2500 Ci/mmol) (Amersham).
4. rNTPs (Pharmacia): ATP, GTP, and CTP stored as 10 mM solution, and UTP as 250 mM solution in nuclease-free distilled water. Store all reagents at –20°C.

2.6. Quantitation of Transcripts

1. Scintillation counter (e.g., Beckman LS 1710).
2. Scintillation vials and scintillation fluid (Amersham).

2.7. Hybridization, Imaging, and Analysis

1. Hybridization buffer: (1 M NaCl, 10 mM Tris-HCl, pH 7.4, 1 mM ethylenediaminetetraacetic acid [EDTA], 0.01% sodium dodecyl sulfate [SDS] [w/v] (*see* **Note 2**).

2. 50–100 fmol radiolabeled transcript.
3. A glass plate of the size of the array when using an array made on glass, a moist chamber (a large plastic or glass-lidded box containing wetted paper towels), and an incubator set to the desired temperature.
4. A hybridization tube and oven used in standard Southern hybridization (e.g., Techne) when using an array made on polypropylene.
5. Esco rubber tubing of OD 1 mm (Sterlin) for use in **Subheading 3.8.**
6. Storage phosphor screen (Fuji or Kodak).
7. PhosphorImager or STORM (Molecular Dynamics).
8. A SUN Solaris work station for image analysis and the computer software xvseq (L. Wand and J. K. Elder, unpublished) (available by anonymous ftp at ftp:// bioch.ox.ac.uk/pub/xvseq.tar.gz).

2.8. Stripping of Arrays

1. Stripping solution: 100 mM sodium carbonate/bicarbonate buffer, pH 10.0, 0.01% SDS [w/v] (*see* **Note 3**).
2. Geiger-Müller counter (Mini-Instruments Ltd.).

3. Methods
3.1. Derivatization of Glass

1. Prepare a mixture of di-isopropylethylamine, glycidoxypropyl trimethoxysilane and xylene (1:17.8:69, v/v/v) in a glass cylinder and completely immerse the glass plates in the mixture. Incubate as shown in **Fig. 3** at 80°C for 9 h.
2. Remove the plates, allow them to cool to room temperature, and wash with ethanol and then with ether by squirting the liquid from a wash bottle.
3. Incubate the plates in hexaethylene glycol containing a catalytic amount of sulfuric acid (approx 25 mL/L) at 80°C for 10 h, with stirring.
4. Remove the plates, allow them to cool to room temperature, and wash with ethanol and ether. Air-dry and store at –20°C.

3.2. Machining of Masks

1. Both stainless steel or aluminium can be used to make diamond-shaped and circular reaction masks. Circular masks are made using a center lathe and diamond-shaped masks using a horizontal milling machine (*see* **Note 4**).
2. To make a diamond shaped mask from metal, hold the work piece at an angle of 45° to the axis of the bed of the milling machine (the diagonal of the diamond running parallel to the axis of the bed).
3. Machine the cavity to the required depth (generally between 0.5–0.75 mm) to create a reaction chamber. Machine the outer lands to a depth of approx 0.5 mm to form the sealing edge (0.3–0.5 mm wide) (*see* **Fig. 4**).
4. Using the smallest possible diameter cutter (approx 1.5 mm), radius the internal corners of the reaction chamber.

5. Finish the sealing edge by polishing flat with successively finer grades of wetted abrasive paper (from approx P600–P1200) and finally with a polishing-grade crocus paper.

6. Drill holes of 1.08 mm diameter for reagent inlet and outlet, respectively, at the bottom and the top of the reaction chamber (in the corners of the diamonds). Inlet and outlet connections to the DNA synthesizer are made using standard 19SWG syringe needles (1.1 mm diameter) with chamfered ends ground off and deburred (*see* **Note 5**).

3.3. Fabrication of Arrays

1. Cut glass or polypropylene to the correct size. The process of making an array is the same when using either glass or polypropylene. Polypropylene has to be mounted on a glass plate, e.g., 3 mm thick soda glass (*see* **Note 6**). The total area covered by an array for N bases using a mask of diagonal or diameter D mm and step size l mm is $N \times l + D$ mm. Two to three milliliters are added to margins to allow easy manipulations.

2. Fix the assembly (*see* **Fig. 4**) to the front of a DNA synthesiser and connect its inlet and outlet to the synthesizer's reagent supply.

3. Program the DNA synthesiser with an appropriate synthesis cycle. A slightly modified cycle is used, for example, the one given in **Table 1**. Also check all the reagent bottles.

4. Enter the sequence (antisense strand) in the 5–3 direction.

5. Mark the first footprint of the reaction mask on the support by placing it against the mask on the assembly in the desired starting position (*see* **Fig. 6**). A knife is used to make notches in polypropylene. A diamond scriber can be used to mark glass.

6. Tighten the plate against the mask with the pressure clamp to produce a seal (*see* **Fig. 4B**). Sufficient pressure is applied to stop leakage (approx 500–800 N force) but not enough to create indentations in the polypropylene surface, which can lead to leakage of reagent from the mask during subsequent synthesis steps.

7. Start the DNA synthesiser to go through the preprogrammed cycle to couple the appropriate nucleotide. The first condensation on the substrate is of base at the 3' end of the sequence.

8. After completion of the step during the interrupt (*see* **Table 1** and **Note 7**), slacken the pressure clamp and move the plate one increment (*see* **Fig. 4B**).

9. Tighten the pressure clamp and start the synthesiser for the next nucleotide in the sequence. Continue the process until the full sequence length is synthesised (*see* **Note 8**).

3.4. Deprotection of Arrays

1. Place the glass or polypropylene array(s) into the HDPE chamber (*see* **Note 9**) and add 30% ammonia into the chamber to cover the array(s) completely.

2. Place the silicon rubber gasket around the rim of the chamber and the stainless steel plate on top of the gasket.

Table 1
Modified Program for an ABI394 DNA/RNA
Synthesizer to Deliver Reagents for One Coupling Cycle

Step no.	Function no.	Function	Step time[a]
1	106	Begin	
2	103	Wait	999.0
3	64	18 to waste	5.0
4	42	18 to column	25.0
5	2	Reverse flush	8.0
6	1	Block flush	5.0
7	101	Phos prep	3.0
8	111	Block vent	2.0
9	58	Tet to waste	1.7
10	34	Tet to column	1.0
11	33	B + tet to column	3.0
12	34	Tet to column	1.0
13	33	B + tet to column	3.0
14	34	Tet to column	1.0
15	33	B + tet to column	3.0
16	34	Tet to column	1.0
17	103	Wait	75.0
18	64	18 to waste	5.0
19	2	Reverse flush	10.0
20	1	Block flush	5.0
21	42	18 to column	15.0
22	2	Reverse flush	10.0
23	63	15 to waste	5.0
24	41	15 to column	15.0
25	64	18 to waste	5.0
26	1	Block flush	5.0
27	103	Wait	20.0
28	2	Reverse flush	10.0
29	1	Block flush	5.0
30	64	18 to waste	5.0
31	42	18 to column	15.0
32	2	Reverse flush	9.0
33	42	18 to column	15.0
34	2	Reverse flush	9.0
35	42	18 to column	15.0
36	2	Reverse flush	9.0
37	42	18 to column	15.0
38	2	Reverse flush	9.0

Table 1 *(continued)*
Modified Program for an ABI394 DNA/RNA
Synthesizer to Deliver Reagents for One Coupling Cycle

Step no.	Function no.	Function	Step time^a
39	1	Block flush	3.0
40	62	14 to waste	5.0
41	40	14 to column	30.0
42	103	Wait	20.0
43	1	Block flush	5.0
44	64	18 to waste	5.0
45	42	18 to column	25.0
46	2	Reverse flush	9.0
47	1	Block flush	3.0
48	107	End	

^{a}Step times are for a diamond-shaped mask having 0.73 mm depth ×
30 mm diagonal and have to be adjusted for each mask.

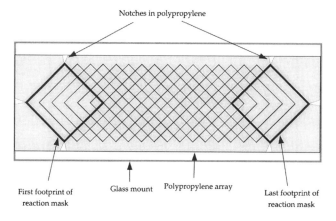

Fig. 6. A completed polypropylene array mounted on a glass plate. The polypropylene has been notched to mark the first and the last footprints of the mask.

3. Place bolts through the metal plate, the gasket and the HDPE chamber, and tighten.
4. Incubate in a water bath at 55°C for 12–18 h in a fume hood.
5. Cool the assembly to 4°C before opening. The arrays are ready to be used in hybridization at this stage.

3.5. Preparing and Quantifying Radiolabeled Transcripts

1. Set an in vitro transcription reaction (20 μL) by adding the following components to a microfuge tube at room temperature.

    a. 5X transcription buffer   4 μL
    b. 100 mM DTT   2 μL
    c. RNAsin®   20 U
    d. 10 mM ATP, GTP, and CTP   1 μL each
    e. 250 mM UTP   1 μL
    f. Template DNA   2–3 μL (*see* **Note 10**)
    g. [α-^{32}P]UTP or [α-^{33}P]UTP   2 μL
    h. T7 or SP6 RNA polymerase   20 U
    i. Total volume   20 μL
    j. Mix and incubate at 37°C for 1 h.

2. Remove 1 μL for quantitation (*see* below).
3. Remove unincorporated label by Sephadex G25 column chromatography (*see* **Note 11**).
4. Save 1 μL of the purified transcript for quantitation (*see* below).
5. Check the integrity of the transcript by denaturing polyacrylamide gel electrophoresis *(10)*.
6. Add 10 μL of the scintillation fluid to the samples saved in **steps 2** and **4**.
7. Mix by vortexing and count the samples in a scintillation counter for 1 min.
8. Calculate the percent incorporation:

$$\% \text{ incorporation} = (\text{incorporated cpm/total cpm}) \times 100$$

9. Calculate the amount of RNA made:

$$\text{Amount of } [\alpha\text{-}^{32}\text{P}]\text{UTP} = (20\ \mu\text{Ci}/3000\ \mu\text{Ci/nmol}) = 6.6 \times 10^{-3} \text{ nmol}$$

$$\text{Amount of cold UTP} = 1\ \mu\text{L} \times 250\ \mu M = 0.250 \text{ nmol}$$

$$\text{Total UTP} = 6.6 \times 10^{-3} + 0.25 = 0.256 \text{ nmol}$$

For a reaction with 50% incorporation, the amount of UTP incorporated

$$= 0.256 \text{ nmol} \div 2 = 0.128 \text{ nmol}$$

Supposing equal incorporation of all four nucleotides, total nucleotides incorporated

$$= 0.128 \text{ nmol} \times 4 = 0.512 \text{ nmol}$$

Amount of full-length transcript

$$= 0.512 \div \text{total length of transcript}$$

3.6. Hybridization to Arrays Made on Polypropylene

1. Place the array in the hybridization tube, coiling it in a spiral.
2. Dilute the radiolabeled transcript in an appropriate volume (10–20 mL depending on the size of the array and the hybridization tube) of hybridization buffer. The mix should cover the array along the length of the tube.

3. Place **items 1** and **2** in the oven at desired temperature for 30 min. Also put approx 100 mL of the hybridization buffer in the oven: this is to be used to wash the array at the end of hybridization.
4. Pour the hybridization mix into the tube containing the array and hybridize for 3–4 h.
5. Remove the hybridization mix. Briefly wash the array with the hybridization buffer from **step 3**, air-dry, cover with cling film and expose to a storage phosphor screen for 16–20 h (*see* **Note 12**).
6. Scan the screen on PhosphorImager or STORM and analyze the image using xvseq (*see* below).

3.7. Hybridization to Arrays Made on Glass

1. Clean the nonarray glass plate with acetone and ethanol to ensure it is grease free and siliconize it by treatment with dimethyl dichlorosilane solution and place it in lidded box. Also place moist paper towel in the box.
2. Dilute the radiolabeled transcript in an appropriate volume of the hybridization buffer (e.g., for an array of 250 × 50 mm use 500–750 mL).
3. Place **items 1** and **2** in an incubator at desired temprature for 30 min. Also put approx 100 mL of the hybridization buffer in the oven: this is to be used to wash the array at the end of hybridization.
4. Using a micropipet, pipet the hybridization mix in a line evenly along the length of the nonarray glass plate, avoiding formation of air bubbles.
5. Starting at one end, carefully place the scanning array (face down) on top of the hybridization mix. The mix will spread out and form a thin film between the two plates. Incubate for 3–4 h.
6. Separate the plates from each other and wash the array plate with hybridization buffer to remove unbound mix. Drain the plate, air-dry, cover with cling film and expose to a storage phosphor screen for 16–20 h.
7. Scan the screen on PhosphorImager or STORM and analyze the image using xvseq (*see* **Subheading 3.9.**).

3.8. Alternative Hybridization Protocol for Glass or Polypropylene

1. Assemble with clips the array plate (or polypropylene array pasted with PhotoMount™ on a glass plate) and the non-array plate, using rubber tubing as spacers on two sides.
2. Dilute the radiolabeled transcript in approx 5–10 mL of hybridization buffer.
3. Follow **Subheading 3.7., step 3**.
4. Inject the hybridization mix into the space between the two plates with a needle and syringe.
5. Incubate the assembly in horizontal position at desired temperature.
6. Follow **Subheading 3.7., steps 6** and **7**.

3.9. Image Analysis

The hybridization images are analyzed using xvseq (*see* **Fig. 7**). This program reads and displays images generated by a PhosphorImager or STORM

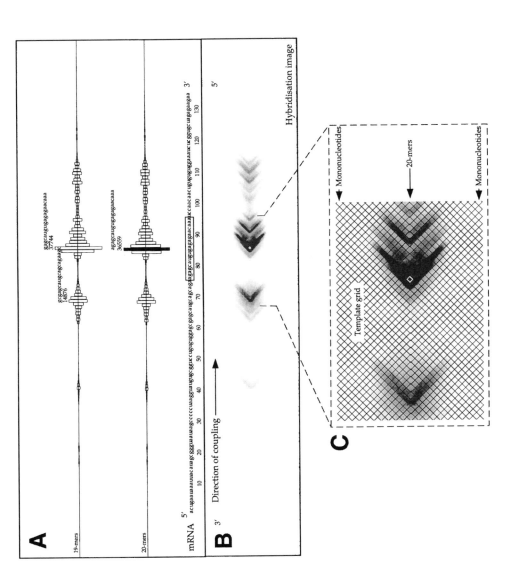

and can also perform standard image manipulation such as scaling, clipping, and rotation. Although visual inspection of an image reveals the results generally, computer-aided analysis is needed to obtain quantitative information about hybridization intensities and the oligonucleotide sequences that generated them. xvseq calculates and displays integrated intensities of the array oligonucleotides, which corresponds to an image cell formed by intersection of overlapping array templates.

The user can specify the template size, shape, and location, step size between successive templates, as well as the sequence that was used to make the array. The template grid is superimposed on the image and the template parameters are adjusted interactively to achieve correct and accurate registration of the grid with the hybridization pattern. It can be difficult to achieve precise registration by reference to the hybridization pattern alone, especially, if the signals at either edge of the array are weak or undetectable. Avoid placing the template grid so that it appears to be registered but is, in fact, misaligned by one or more template steps. Registration can be aided by the use of fixed reference points on an array such as those shown in **Fig. 6**.

3.10. Stripping of Arrays

1. The arrays can be used several times. To strip, heat an appropriate volume of the stripping solution to 90°C in a glass beaker.
2. Immerse the array in the hot stripping solution and stir for 1–2 min.

Fig. 7. *(previous page)* Display and analysis of hybridisation results using xvseq. (**A**) The results of the integrated pixel values for 19-mers and 20-mers are displayed as paired histograms; the two values are for the areas above and below the center line. The program can display all values from monomers to the longest oligonucleotides on the array and also includes tools to aid the interpretation of the large amounts of data in the figure. For example, clicking on a bar in the histogram (e.g., filled bar) highlights the corresponding cell (oligonucleotide) in the array and the region in the sequence (boxed sequence). Clicking on a cell in the array highlights the corresponding region in the sequence (displayed on the relevant histogram), the relevant bar in the histogram and the integrated pixel values of the two cells. The values are of hybridization intensities of the oligonucleotides on a single array and should not be used directly as comparison between different arrays. (**B**) Hybridization of a ^{32}P-labeled mRNA to an array of complementary oligonucleotides. The longest oligonucleotides on this array are 20-mers. The sequence of the region of mRNA complementary to the array is written to a text file and is loaded before starting the integration process. A template grid of a series of overlapping diamonds, corresponding in number to the sequence length, is placed over the image to generate cells containing individual oligonucleotide sequences. (**C**) A part of the image magnified with a template grid overlay.

3. Remove the array and monitor with a Geiger counter to confirm that most of the radiolabel has been removed. Repeat **steps 1** and **2** if radioactivity on the surface of the array is still detectable.
4. Allow the array to cool down to room temperature and wash it thoroughly with nuclease-free distilled water, 70% (v/v) ethanol and finally with absolute ethanol.
5. Air-dry and store the array at –20°C until future use.

4. Notes

1. The choice of array substrate material and attachment chemistry is important for making high-quality arrays. A flat, impermeable surface is required for *in situ* synthesis of arrays. Glass has a number of favorable qualities, including its wide availability, smooth surface, transparency, chemical stability, and compatibility with the use of both radiolabeled or fluorescence-labeled nucleic acids targets. Glass is chemically derivatised as described in **Subheading 3.** to produce a hexaethylene glycol linker that has a terminal –OH group that allows condensation of nucleotide phosphoramidites *(11)*. Similarly, polypropylene also has favourable physical and chemical properties. Polypropylene is aminated to produce amine groups *(12)* that also allow synthesis to oligonucleotides using standard CE nucleotide phosphoramidites.

   For the array fabrication method described here, it is important that a tight seal is formed between the substrate material and the reaction mask. Metals form tight seal with polypropylene but not with glass. PTFE seals well against both glass and polypropylene.
2. We use 1 M NaCl routinely. Alternative buffers are: 1 M NaCl, 5–10 mM MgCl$_2$, 10 mM Tris-HCl, pH 7.4, 1 mM EDTA, 0.01% SDS (w/v), and 150 mM NaCl, 10 mM MgCl$_2$, 10 mM Tris-HCl, pH 7.4, 1 mM EDTA, 0.01% SDS (w/v).
3. Addition of more than 0.01% SDS can damage arrays.
4. Circular masks can also be made from PTFE (Teflon). Diamond-shaped masks are more difficult to make with PTFE by the machining process but can be made by pressure molding in a hydraulic press (approx 150 ton force) using a premachined die.
5. Holes should be made as close as reasonably possible to the sealing edge without damaging it. Care must be taken to deburr fully the holes at the point of entry into the reaction chamber. For PTFE masks the holes should be 1.0 mm diameter which make virtually 100% leak-tight seal. In the case of metal masks, the 0.02-mm interface indicated above also provides a leak-tight seal without the use of any additional sealer. Care must be taken not to insert the end of the syringe needle into the reaction chamber void.
6. Unlike glass, polypropylene is not rigid and thus needs to be mounted on a solid, flat surface for its precise movement against the reaction mask during synthesis. Even mounting of polypropylene on glass is important to produce a good seal between the sealing edge of the reaction mask and the polypropylene surface. Glass used must be clean and free from dust particles because they can cause bulging of the polypropylene, which can hinder the formation of a proper seal. A

Table 2
Bacteriophage Promoter and Leader Sequences[a]

T7	5' taa tac gac tca cta ta *ggg cga*
(or)	5' taa tac gac tca cta ta *ggg aga*
SP6	5' att tag gtg aca cta ta *gaa tac*

[a]The hexa-nucleotide leader sequence *(in italics)* appears in the transcript.

very thin layer of PhotoMount™ (3M Inc.), which can be used to paste polypropylene to glass, should be sprayed on glass and not polypropylene.

7. At the start of each synthesis cycle, an interrupt step can be introduced to halt the process at the first step of the next nucleotide condensation cycle to allow the operator to move the plate and restart the program. Alternatively, a long wait step at the beginning of the program can be introduced (*see* **Table 1**) if the operator does not wish to use the interrupt step. The operator is also advised to consult the user's manual for the DNA synthesizer.

8. With the use of standard phosphoramidites in the synthesis, the oligonucleotides are attached to the solid support at their 3' ends. Reverse phosphoramidites can be used to make oligonucleotides that are attached at their 5' ends.

   Iodine is used as an oxidizing agent. At lower temperatures it will take longer to reach the top of the reaction cell. Iodine can also be replaced with sulfurizing agent (Cruachem) to make arrays of phosphorothioate oligonucleotides.

9. When using the standard phosphoramidites, the exocyclic amines of the bases are protected chemically to prevent side reactions during synthesis. These protecting groups need to be removed from the coupled bases before hybridization.

   Before deprotection, detach the polypropylene arrays from glass by peeling from one end. PhotoMount can be removed with ethanol, acetone, or dichloromethane.

10. An internally radiolabeled RNA is used as target to hybridize to a scanning array, which is generated by in vitro transcription, carried out in the presence of $[\alpha\text{-}^{32}P]UTP$ or $[\alpha\text{-}^{33}P]UTP$ (or $[\alpha\text{-}^{32}P]CTP$) using an appropriate DNA template. A plasmid containing the desired DNA fragment under the transcriptional control of a T7 or SP6 promoter (such as pGEM: Promega) can be used as template. The plasmid is linearized with an appropriate restriction endonuclease to produce transcripts of defined length without contaminating vector sequence. Alternatively, a template with T7 or SP6 RNA promoter can also be generated using the polymerase chain reaction: primers are used to amplify the required fragment from a plasmid, genomic DNA or cDNA, such that the sense primer has a T7 or SP6 promoter leader sequence (*see* **Table 2**) added at the 5' end.

11. Sephadex G25 columns are available from several commercial suppliers including Promega and Pharmacia. Spin columns made in-house, as described in (**10**), can also be used.

12. For hybridization below 37°C, care must be taken not to touch the plates because this can lead to melting of short duplexes. For hybridization below room tem-

perature, the cling film and the phosphor screen must be cooled to hybridization temperature and exposed at the same temperature.

Acknowledgments

The authors would like to thank U. Maskos, J. K. Elder, L. Wang, S. C. Case-Green, M. J. Johnson, K. Mir, N. Milner, L. Baudouin, and J. Williams for their assistance in the development of methods described in this chapter.

References

1. Southern, E. M., Case-Green, S. C., Elder, J. K., Johnson, M. Mir, K. U., Wang, L., and Williams, J. C. (1994) Arrays of complementary oligonucleotides for analysing the hybridisation behaviour of nucleic acids. *Nucleic Acids Res.* **22,** 1368–1373.
2. Milner, N., Mir, K. U., and Southern, E. M. (1997) Selecting effective antisense reagents on combinatorial oligonucleotide arrays. *Nat. Biotechnol.* **15,** 537–541.
3. Mir, K. U. and Southern, E. M. (1999) Determining the influence of structure on hybridisation using oligonucleotide arrays. *Nat. Biotechnol.* **17,** 788–792.
4. Zuker, M. (1989) On finding all suboptimal foldings of an RNA molecule. *Science* **244,** 48–52.
5. Dumas, J.-P. and Ninio, J. (1982) Efficient algorithms for folding and comparing nucleic acid sequences. *Nucleic Acids Res.* **10,** 197–206.
6. Ho, S. P., Bao, Y., Lesher, T., Melhotra, R., Ma, L. Y., Fulharty, S. J., and Sakai, R. R. (1998) Mapping of RNA accessible sites for antisense experiments with oligonucleotide libraries. *Nat. Biotechnol.* **16,** 59–63.
7. Sohail, M., Akhtar, S., and Southern, E. M. (1999) The folding of large RNAs studied by hybridisation to arrays of complementary oligonucleotides. *RNA* **5,** 646–655.
8. Branch, A. D. (1998) Antisense drug discovery: Can cell-free screens speed the process? *Antisense Nucleic Acid Drug Develop.* **8,** 249–254.
9. Sohail, M. and Southern, E. M. (2000) Selecting optimal antisense reagents. *Adv. Drug Deliv. Rev.*, in press.
10. Sambrook, J., Fritsch, E. F., and Maniatis, T. (1989) *Molecular Cloning: A Laboratory Manual.* Cold Spring Harbor Laboratory Press, Cold Spring Harbor, NY.
11. Maskos, U. and Southern, E. M. (1992) Oligonucleotide hybridisation on glass supports: a novel linker for oligonucleotide synthesis and hybridisation properties of oligonucleotides synthesised *in situ. Nucleic Acids Res.* **20,** 1679–1684.
12. Matson, R. S., Rampal, J. B., and Coassin, P. J. (1994) Biopolymer synthesis on polypropylene supports. I. Oligonucleotides. *Anal. Biochem.* **217,** 306–310.

13

Low-Resolution Typing of *HLA-DQA1* Using DNA Microarray

Sarah H. Haddock, Christine Quartararo, Patrick Cooley, and Dat D. Dao

1. Introduction

Over the last few decades, typing for the human leukocyte antigen (HLA) has become critical for bone marrow transplants. Research has also indicated that individuals with certain allelic genotypes may be at a higher risk for developing diseases affecting the immune system, and therefore rapid tests to identify the numerous allelic differences are needed. Such tests must allow for the detection of the various alleles present in a 4-Mbp region on the short arm of chromosome 6. The area comprises three class regions: class I, which includes *HLA-A*, *HLA-B*, and *HLA-C* genes; class II, containing the *HLA-D* genes; and class III, composed of numerous protein coding genes responsible for a range of functions *(1,2)*.

Class II genes are composed of both α- and β-chains in the *HLA-D* region, consisting of *DP*, *DQ*, and *DR* genes. This area is responsible for more than 500 different alleles with the greatest concentration in the *DRB1* gene. The *DRB* region also varies from other areas of the *HLA* complex in that either one or two functional genes may be present *(3)*. This gene is one of the prime areas under investigation for allelic identification by DNA sequencing for the determination of individuals with matching types.

An ideal area to construct a model for rapid typing of *HLA* is located in the *DQA1* region of the *HLA-D* gene near the centromere. This area has approx 15 different allelic possibilities responsible for assisting in the distinction of intruder cells from the body's own cells *(4)*. Therefore, the site is one of many locations where the donor and recipient are required to have identical alleles in order to avoid rejection of the transplant material. Yet, the *DQA1* locus tends

From: *Methods in Molecular Biology, vol. 170: DNA Arrays: Methods and Protocols*
Edited by: J. B. Rampal © Humana Press Inc., Totowa, NJ

to be an ideal area of basic human identification rather than one used to type transplant material. The region is ideal for study because the number of alleles is substantially less compared to that of other areas, which can provide simplicity and usefulness in the development of a model.

The need for a rapid and accurate test to distinguish one allele from another has become our primary focus. The test we propose utilizes short oligonucleotide probes of 12 bases in length, attached to a glass slide. Each probe represents a different low-resolution allele in the *DQA1* region. Biotinylated polymerase chain reaction (PCR) products are allowed to hybridize to the prefabricated glass slides, and hybridization signals are detected using enzyme-labeled fluorescence (ELF) with an aid of a charge-coupled device (CCD) camera.

2. Materials

2.1. DNA Samples

The standardized DNA Reference Panel for *HLA* class II was obtained from the University of California at Los Angeles (UCLA) Tissue Typing Laboratory. Samples representing the *DQA1* class of alleles were used in the selection of the optimal probes and then blinded and used for detection purposes (*see* **Note 1**).

2.2. Oligonucleotide Probe Synthesis

Amine-modified probes were produced by means of a standard phosphoramidite and a segmented synthesis strategy for the simultaneous synthesis of a large number of oligonucleotides (Sigma-Genosys, The Woodlands, TX). **Figures 1** and **2** show the location and sequence of optimal probes for alleles 01–06 along the amplified region of target DNA. Oligo Software (Molecular Biology Insights, Cascade, CO) was used for the probe design (*see* **Note 2**).

2.3. Preparation of Target

1. 10 m*M* Tris-HCl, pH 8.0, 50 m*M* KCl, 0.01% gelatin.
2. MgCl$_2$.
3. AmpliTaq Gold DNA Polymerase (PE Biosystems, Foster City, CA).
4. dNTPs (Amersham, Piscataway, NJ).
5. Biotinylated primers with the following sequences (Sigma-Genosys):
   a. Forward primer: 5'CACCCATGAATTTGATGGAG 3'.
   b. Reverse primer: 5'TCATTGGTAGCAGCGGTAGAGTTG 3'.
6. DNA samples (UCLA, Los Angeles, CA).
7. High-performance liquid chromatography (HPLC) H$_2$O (Fisher, Pittsburgh, PA).

2.4. Attachment

1. Frosted glass microscope slides (Gold Seal, Portsmouth, NH).
2. Printing of probes: MicroFab Technologies, Plano, TX.

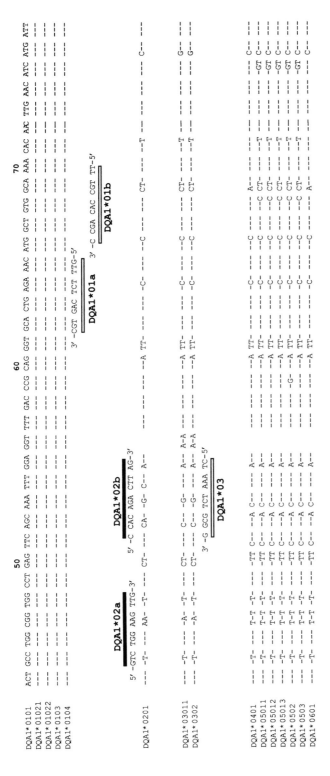

Fig. 1. Low-resolution probes for the typing of *HLA DQA1*01*, *DQA1*02*, and *DQA1*03* alleles: a region of the *HLA-DQA1* gene showing variations in the sequence for each allele. Probe sequences are shown for alleles 01, 02, and 03 by either a solid bar (sense probe) or open bar (antisense probe) over the area where the probe is located.

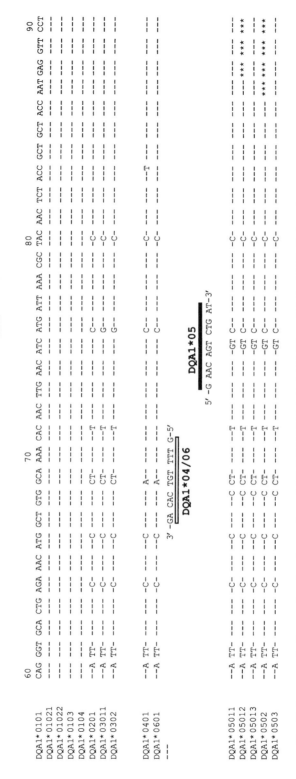

Fig. 2. Low-resolution probes for the typing of *HLA DQA1*04*, *DQA1*05*, and *DQA1*06* alleles: the gene sequence from a region of the *HLA-DQA1* with unique sequence shown for each allele. A solid bar (sense probe) or open bar (antisense probe) is placed above the probe sequences for the 04/06 and 05 alleles.

2.5. Hybridization

1. Slides with probe array attached.
2. 6X saline sodium citrate (SSC) (0.9 M sodium chloride, 0.09 M sodium citrate), 10% sodium dodecyl sulfate (SDS).
3. Amplified target DNA (5 µL).
4. Corning 24 × 50 mm cover glasses (Fisher).
5. 3.3 M tetramethyl ammonium chloride (USB, Cleveland, OH).

2.6. Detection of Fluorescence

1. Avidx-Ap Streptavidin (Tropix, Bedford, MA) mixed into a solution (1:1500 dilution in phosphate-buffered saline (PBS), 0.2% bovine serum albumin (BSA), 0.1% Tween-20).
2. 2X SSC/0.1% SDS solution.
3. ELF substrate (cat. no. 6601, Molecular Probes, Eugene, OR).
4. Ultraviolet (UV) light with a CCD camera (SpectraSource, Westlake Village, CA).

3. Methods
3.1. Oligonucleotide Probe Design

Oligonucleotide probes were designed using computer software developed in C++. The program determined the sequence of probes for the selected allele by looking for an area of the sequence that matched the order for the selected allele but for no other group. Probes were then chosen by taking the desired probe length, suitable G/C content, and the ideal placement of the mismatched base(s) near the center of the region in consideration. The values of free energy were also examined when developing optimal probes by employing the use of the Oligo Software program (*see* **Note 2**).

3.2. Oligonucleotide Attachment

MicroFab Technologies performed attachment of probes by means of drop-on-demand ink jet devices to create DNA microarrays. These devices comprise a glass capillary and lead-zirconate-titante (PZT) cylinder and were used to deliver picoliter volumes of the oligonucleotide probes onto the glass slides (*see* Chapter 7). A 5 µ*M* concentration of amine-modified probe solutions was used for attachment onto glass slides.

3.3. Preparation of Target

A 189-bp region of the *HLA-D* gene was amplified using a total volume of 50 µL consisting of 10 m*M* Tris-HCl, pH 8.0; 50 m*M* KCl; 0.01% gelatin; 100 µ*M* dNTPs; 1.5 m*M* MgCl$_2$; 20 pmol of each primer; 2.5 U of AmpliTaq Gold DNA Polymerase; and 100 ng of the DNA sample. The amplification was carried out with a 10-min activation time at 94°C, followed by 30 cycles at 94°C for 30 s, 54°C for 1 min, and 72°C for 30 s.

3.4. Hybridization Conditions

Hybridizations were performed in a 50-µL solution consisting of 6X SSC, 0.3% SDS, and 5 µL of the amplified DNA. Heat-denatured target DNA was added to a glass slide prewetted in 2X SSC and 0.1% SDS. The glass slide was covered with a 24 × 50 mm glass cover slip and then placed into a moist chamber for 30 min at room temperature. The slides were washed in 3.3 *M* tetramethyl ammonium chloride (TMACl) for 15 min at room temperature (*see* **Note 3**).

3.5. Detection of Fluorescence

The slides were then incubated for 2 h in a moist chamber with 50 µL of streptavidin solution (1:1500 dilution in PBS/0.2% BSA/0.1% Tween-20). After two 10-min washes in a 2X SSC/0.1% SDS solution, 50 µL of ELF substrate (mix components A and B; 1:19 respectively) was added to the glass slide and covered with another cover glass. The slides were placed in a dark environment overnight and then viewed under UV light with a CCD camera.

To distinguish all alleles found in the *DQA1* region of the *HLA* gene, two different microarrays were used. The first array used included alleles 01, 02, and 03, which had the ability to detect each allele as well as a combination of any two alleles. This array is shown in **Fig. 3** and includes two probes each for 01 and 02 and one probe for 03 as well as images depicting signals for the alleles. In sample A, probes number 1 and 2 illuminated under UV light, thus determining that this sample must represent allele 01. Positive signals were emitted by probes number *DQA1*02a* and *DQA1*02b* in sample B, indicating the presense of the 02 allele. Sample C hybridized to probe number *DQA1*03*, specifying the 03 allele. Probes representing alleles 01 (*DQA1*01a* and *DQA1*01b*) and 03 (*DQA1*03*) were viewed under UV light in sample D, exhibiting an individual with a heterozygous genome.

The second array included probes for alleles 04/06 and 05. Thus, this array was able to detect all of the low-resolution alleles as well as combinations of alleles for individuals having a heterozygous genotype; this is exhibited in **Fig. 4**. Sample A hybridized to probe *DQA1*01a* and *DQA1*01b*, once again showing that the sample contains allele 01. Allele 03 was visible by viewing positive signals for *DQA1*03* in sample E. Probe *DQA1*04/06* illuminated under UV light in sample F, indicating that allele 04 or 06 was present in this sample. Sample G was identified to possess allele 05 by hybridizing to probe *DQA1*05* in array 2. Both allele 01 and allele 04 were found to be contained in sample H; this is evident by probe *DQA1*01a* and *DQA1*04/06* fluorescing under UV light (*see* **Note 4**).

We have successfully typed all low-resolution alleles using a 189-bp region of the *HLA-DQA1* gene by observing easily recognizable positive hybridiza-

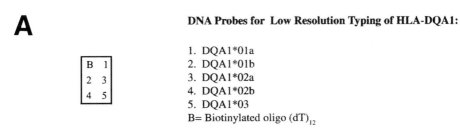

A **DNA Probes for Low Resolution Typing of HLA-DQA1:**

B	1
2	3
4	5

1. DQA1*01a
2. DQA1*01b
3. DQA1*02a
4. DQA1*02b
5. DQA1*03
B= Biotinylated oligo (dT)$_{12}$

B

Sample	Hybridization Pattern	Probe Signal	Allele
A		1 2	*01
B		3 4	*02
C		5	*03
D		1 2 5	*01 *03

Fig. 3. Low-resolution typing of *HLA-DQA1* alleles 01, 02, and 03 (**A**) array 1: DNA microarray illustrating positions of probes for alleles 01, 02, and 03 in the *HLA-DQA1* gene. On the right of the microarray is probe nomenclature for the corresponding probe numbers on the DNA microarray. (**B**) Samples A–D were tested on the above array in a blinded fashion. Alleles found by DNA chip analysis are displayed by positive probe signals. The blinded samples were then confirmed by referring to the prepared key and the information provided in the standard reference material. All photographs were taken with the use of a CCD camera under a UV light source.

tions patterns (*see* **Note 5**). In **Figs. 3** and **4**, detection of both homozygous and heterozygous samples was determined by viewing one or more probes fluorescing under UV light. The importance of this type of array is that it encom-

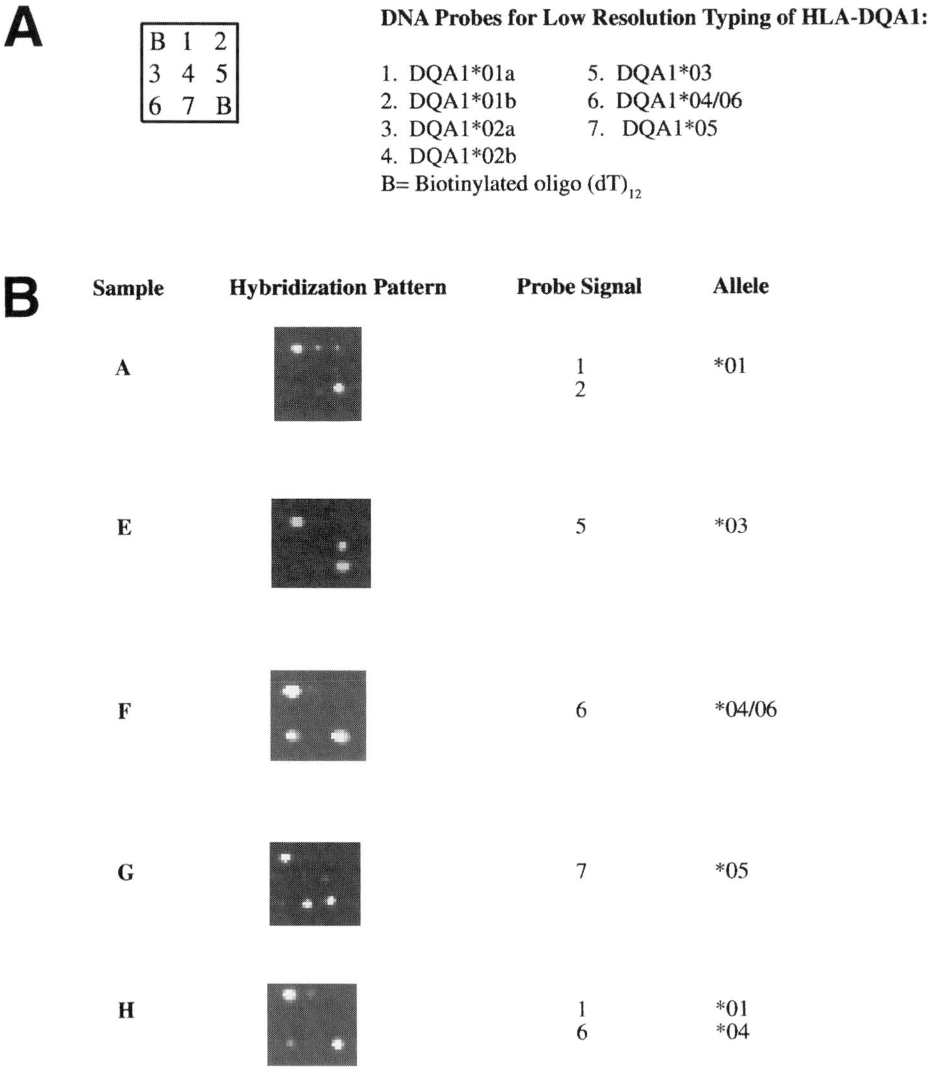

A

B	1	2
3	4	5
6	7	B

DNA Probes for Low Resolution Typing of HLA-DQA1:

1. DQA1*01a   5. DQA1*03
2. DQA1*01b   6. DQA1*04/06
3. DQA1*02a   7. DQA1*05
4. DQA1*02b

B= Biotinylated oligo (dT)$_{12}$

B

Sample	Hybridization Pattern	Probe Signal	Allele
A		1 2	*01
E		5	*03
F		6	*04/06
G		7	*05
H		1 6	*01 *04

Fig. 4. Low-resolution typing of *HLA-DQA1* alleles 01–06 (**A**) array 2: DNA microarray depicting positions of probes for alleles 01, 02, 03, 04/06, and 05. Shown to the right is probe nomenclature for the corresponding probe numbers on the DNA microarray. (**B**) Samples A and E–H hybridization results from slides prepared with the microarray above. The probe signals are viewed directly with the use of a CCD camera under a UV light source. A key developed correlating to the UCLA nomenclature from the standard reference material was used to confirm the results.

passes the use of short oligonucleotide probes providing a high level of discrimination between alleles, which can allow this type of array to serve as a model for typing other important gene regions in the *HLA* complex.

4. Notes

1. Each DNA sample was labeled with a letter or letters that corresponded to the key provided in the standard reference material. Then the experiment was run and the results were compared to the key to verify that the *HLA* typing of the sample was the same.

2. In this study, oligonucleotide probes were designed to capture the unique sequence of a particular allele. Several criteria were taken into consideration during the design and selection of the probes, such as optimal probe length of 12 bases to allow hybridizations to be performed at room temperature; ideal placement of the mismatch base(s) near the center of the sequence; suitable G/C content ranging from 42 to 58%; and the values of free energy to be greater than –3.1 ΔG, so that the probe duplex formations would be kept to a minimum. Originally, probes were designed manually by comparing the sequence of each allele to one another and selecting a region where each allele could be distinguished from one another. This was done without much success and therefore other options were explored. One option was to design computer software that would implement each parameter listed above as well as search out the ideal placement of the probe along the sequence of the gene. In each instance this proved to be the best method to use in the selection of the probes to distinguish the alleles from one another on the microarrays. Several probes were calculated by the software and tested to identify a probe that emitted the strongest signal without crosshybridizing with samples containing different alleles. A combination of sense and antisense sequences was used to formulate the microarray for low-resolution *HLA-DQA1* typing. Probes were detected for each allele, with the exception of the 04/06 probe (*see* **Figs. 1** and **2**). This probe detects both the sequence of the 04 allele and that of the 06 allele, because at no point in the amplified region of DNA were the sequences different from one another. Therefore, this probe had to include both alleles (**Fig. 2**).

3. Experiments to optimize hybridization conditions for the identification of low-resolution alleles in the *HLA-DQA1* gene region required various hybridization times, washing times, and temperatures. Hybridizations were performed at 15, 20, 30, 45, and 60 min at room temperature with the optimal time at 30 min in combination with TMACl wash times and temperatures, ranging from room temperature to 42°C and times ranging from 10 to 25 min. Washing slides for 15 min at room temperature in 3.3 *M* TMACl led to the least crosshybridization and the strongest signals.

4. Two arrays were used in this study. The first array (**Fig. 3**) allowed for the verification of the working capabilities of the system. The second array (**Fig. 4**) was employed once, and optimal probes for alleles 04/06 and 05 were discovered, thus forming an array that includes all low-resolution alleles for the *DQA1* region.

5. Forming a successful working array for the *HLA-DQA1* region as well as attaining a computer program with the ability to select probes for use in the detection process enables this system to serve as a model for other genes within the *HLA* region. The complexity of such an array is greatly diminished, because only one

probe is needed for each allele, and thus there is ease in interpreting the results. The *HLA-DRB1* gene is one of the most valuable areas in typing individuals for *HLA*. This region has more than 254 different alleles *(5)*. Therefore, further development of the software used to design probes containing *HLA-DR* alleles is needed in order to create arrays for *HLA* typing.

Acknowledgments

We wish to thank Melissa Drysdale for her expert technical assistance, Bob Matson of Beckman Coulter (Fullerton, CA) for the introduction of ELF to be used as a detection system in our assay, as well as Fan Gong and Lori Dalton for the development and refinement of the computer software program used for the design of oligonucleotide probes. We also thank Bernie Kosicki, Jim Caunt, Brad Felton, and colleagues at the Massachusetts Institute of Technology (MIT) Lincoln Laboratory (Lexington, MA) for their consultation in the design of the UV source CCD camera reader. In addition, we wish to thank Rick Staub of Laboratories for Genetic Services for his interest in the development of the HLA microarray chip. We gratefully acknowledge support received through Cooperative Agreement No. 70NANB3H1364 between The Genosensor Consortium (Beckman Coulter, Baylor College of Medicine, Genometrix, Genosys Biotechnologies, Laboratories for Genetic Services, MicroFab Technologies, MIT Lincoln Laboratory, and Houston Advanced Research Center), and the U.S. Department of Commerce/National Institutes of Standards and Technology/Advanced Technology Program.

References

1. Trowsdale, J. (1996). Molecular genetics of HLA class 1 and Class II regions, in *HLA and MHC Genes, Molecules, and Function* (Browning, M. and McMichael, A., eds.), BIOS Scientific Publishers, Oxford, pp. 23–38.
2. Aguado, B., Milner, C. M., and Campbell, R. D. (1996) Genes of the MHC class III region and the function of the proteins they encode, in *HLA and MHC Genes, Molecules, and Function.* (Browning, M., and McMichael, A., eds.), BIOS Scientific Publishers, Oxford, pp. 39–75.
3. Begovich, A. B. and Erlich, H. A. (1995) HLA typing for bone marrow transplantation. *J. Amer. Med. Assoc.* **273,** 586–591.
4. Bodmer, J. G., Marsh, S. G. E., Albert, E. D., Bodmer, W. F., Bontrop, R. E., Charron, D., et al. (1995) Nomenclature for factors of the HLA system. *Tissue Antigens* **46,** 1–18.
5. Robinson, J. (2000) IMGT/HLA Database Statistics <http://www3.ebi.ac.uk/Services/imgt/hla/cgi-bin/statistics.cgi>

14

Gene Expression Analysis on Medium-Density Oligonucleotide Arrays[1]

Ralph Sinibaldi, Catherine O'Connell, Chris Seidel, and Henry Rodriguez

1. Introduction

As the Human Genome Project continues toward its goal of characterizing the genome of human and selected model organisms through complete mapping and sequencing of their DNA, unique opportunities are becoming available for studying genetic variation in humans and its relationship with disease risk and aging. Consequently, new techniques have been invented to rapidly screen genes for biological information.

Over the past few years, DNA microarrays have received considerable attention from both researchers and the public. DNA arrays represent a blossoming field, estimated to be worth $40 million a year and expected to grow 10 times that over the next 5 yr. DNA arrays are the latest molecular biology technique to utilize nucleic acid hybridization as its basis. Two of the most common uses of the DNA arrays are genetic analysis and the analysis of gene expression. Genetic analysis includes procedures for genotyping, single nucleotide polymorphism detection, strain identification, and various other procedures. Analysis of gene expression can include assessing expression levels of small sets of genes to whole-genome expression monitoring. Depending on the design of the arrayed DNAs, other expression phenomena such as differential RNA splicing can also be examined and analyzed on expression arrays. This

---

[1]Certain commercial equipment or materials are identified in this chapter to adequately specify experimental procedures. Such identification does not imply endorsement by the National Institute of Standards and Technology, nor does it imply that the materials or equipment identified are necessarily the best available for the purpose.

From: *Methods in Molecular Biology, vol. 170: DNA Arrays: Methods and Protocols*
Edited by: J. B. Rampal © Humana Press Inc., Totowa, NJ

chapter focuses on using medium-density oligonucleotide DNA arrays for the analysis of gene expression.

Early studies of DNA melting and reformation were conducted in aqueous solutions and yielded important information about the dependence of T_m on the G+C composition and salt concentration, as well as information on the dependence of the rate of reassociation on the sequence complexity of the nucleic acid. The introduction of solid supports for DNA hybridization/reassociations greatly broadened the range of applications of nucleic acid hybridizations, and provided the basis for solid-based methods being used today. It was demonstrated that when double-stranded DNA is denatured, the resultant single-stranded DNA binds strongly to nitrocellulose membranes in a manner that minimizes the two strands reassociating with each other, but allows the hybridization to complementary RNA *(1)*. This method was used to measure the number of copies of repeated genes such as *rRNA* genes *(2)* and to measure whether specific genes were underreplicated during the replication process used in forming polytene chromosomes *(3)*. Dot blotting and dot hybridization *(4)* evolved out of the filter hybridization technique and provided the fundamental concept for today's DNA arrays. The difference between dot blots and DNA arrays lies in the use of a nonporous rigid solid support such as glass, which has its advantages over a porous membrane. The membranes require much larger volumes of hybridization solutions to hybridize to the immobilized DNA in the porous substrate. The glass support requires very small volumes of hybridization solution, and more rapid rates of hybridization are evident when compared to filter hybridizations. The glass support also facilitates the washing step of the hybridization and provides a matrix of very low inherent fluorescence so that fluorescently labeled probes can be effectively utilized in the hybridization. Furthermore, the glass support provides a substrate to which DNA or oligonucleotides can be covalently, and thus stably attached.

The two most common types of DNA arrays are those in which the DNA (in the form of a single-stranded oligonucleotide) is actually synthesized *in situ* *(5)* and those in which the DNA (usually in the form of a cDNA or full length open reading frame [ORF]) is postsynthetically attached to a solid support *(6)*. The first type is quite useful for genetic analysis, which utilizes relatively short oligonucleotides, but can be used for expression analysis by using many short oligonucleotides to cover a gene. The length of the synthesized oligonucleotide is limited by the relatively poor coupling efficiency of *in situ* DNA synthesis. The second process usually involves the making of full-length ORFs or cDNAs and postsynthetically printing them on 2.54 × 7.62 cm poly-L-lysine-coated microscope slides. The latter type of arrays is used for the analysis of gene expression but has certain limitations. The first is that ORFs are extremely variable in their length and T_m, making the hybridization signal relatively uneven across the array. Unlike conventional Northern analysis, DNA array experiments make use of a competitive hybridization of two differentially labeled probes for a DNA spot on the array. The cDNA made from the control RNA (or treatment/tissue type 1/or wild type) is labeled with one fluorochrome, whereas the cDNA made from the experimental RNA (or treatment/tissue type 2/or mutant) is labeled with a contrasting fluorochrome. Both labeled cDNAs are hybridized to the same array, and the results are tallied by comparing the ratios of the two fluorescent emissions of the fluorochrome containing cDNAs hybridized to each of the printed DNAs. A fluorescent scanner scans the fluorescent pattern of each fluorochrome, and the two patterns can be overlaid to assess which genes have been upregulated or downregulated by the experimental treatment.

A second and probably more important limitation of the ORF-based DNA arrays is the issue of cross-hybridization of related or overlapping genes. Proteins from genes may have common features (i.e., adenosine triphosphate binding sites) and can have some degree of sequence identity or homology with other genes. Most organisms contain a number of genes in gene families, and in many cases these genes have a great degree of sequence identity to each other and can only be distinguished from each other by designing and using gene-specific hybridization probes. A third limitation is that many organisms have overlapping genes where one gene or ORF is on one strand of the DNA duplex and another gene or ORF is found on the complementary strand of the DNA duplex. For example, in *Saccharomyces cerevisae*, 728 of the approx 6000 genes have 100% sequence identity over a distance of 101 nucleotides. Consequently, the expression levels of the genes cannot be accurately measured using full-length ORF-based DNA arrays. Last, many organisms employ alternative RNA splicing of genes in response to differentiation and other signals, and these alternative forms of gene expression cannot be distinguished on ORF-based arrays.

Many of these limitations of the ORF-based arrays are addressed by designing what is called an *optimized array*. These arrays are based on the design of sequence–optimized DNAs to be arrayed. We employ a rational design to select 70-mers (or DNA sequences) that are sequence optimized and normalized for hybridization temperature. The 70-mers are designed to be complementary to sequences near the 3' ends of genes. In addition, because of the uniform size, we are able to print the oligos at a normalized concentration so that every spot contains a consistent amount of nucleic acid. Hybridization-normalized DNA sequences can be designed to be ± 2°C of each other, thus assuring consistent hybridization for all the DNA sequences on the array. Sequence optimization is a design process in which we minimize cross-hybridization and overlapping gene hybridization by choosing sequences in the gene that do not cross-hybridize with other genes. The oligonucleotides can also be designed to detect alternatively spliced genes. The criticism of the oligonucleotide approach for the analysis of gene expression is that it is not very sensitive and multiple oligonucleotides to cover a gene must be used. This is true with shorter oligonucleotides (15- to 20-mer), but the 70-mers do yield good sensitivity and offer the great specificity of shorter sequences—a better alternative. We compared the sensitivity of the 35-, 50-, 70-, and 90-mer for detecting highly expressed genes and genes expressed at moderate or low levels in yeast, and the 70-mer performed the best. We believe that 70-mer oligo-based arrays offer an accurate and cost-efficient way to monitor gene expression.

The majority of the actual slide making, printing, hybridization, probe making, and posthybridization methods given herein are the same or only slightly modified versions of the protocols found on the Stanford University Web site (http://cmgm.stanford.edu/pbrown/protocols/). The changes made to the Stanford protocols are to accommodate oligonucleotides as the probes on the glass slide.

2. Materials

2.1. Preparation of Slides

1. Glass microscope plain slides 2.54 × 7.62 cm, 1.0 mm thick (Gold Seal, product no. 3010; Fisher, Hampton, NH).
2. Slide rack (product no. 121; Shandon Lipshaw). Each rack holds 30 slides.
3. Slide chamber (product no. 121; Shandon Lipshaw). Each chamber holds 350 mL.
4. Double-distilled water (ddH$_2$O).
5. NaOH pellets.
6. 95% Ethanol.
7. Polylysine solution: 70 mL of poly-L-lysine (Sigma, St. Louis, MO) + 70 mL of tissue culture 1X PBS (Life Technologies, Rockville, MD) in 560 mL of water. Use a plastic graduated cylinder and beaker.
8. Tissue culture 1X PBS (Life Technologies).
9. Vacuum oven (Napco model no. 5831; Fisher).
10. One plastic slide box (product no. 48443-806; VWR Scientific, West Chester, PA).
11. Cleaning solution: Dissolve 70 g of NaOH in 300 mL of ddH$_2$O. Add 420 mL of cold 95% ethanol. Total volume is 700 mL (= 2 × 350 mL). Stir until completely mixed. If the solution remains cloudy, add ddH$_2$O until clear.
12. Orbital shaker (model no. Innova 2000; New Brunswick Scientific, Edison, NJ).
13. Centrifuge (model no. Allegra 6R; Beckman Coulter, Fullerton, CA).

2.2. Arraying of Oligonucleotides

1. Nucleic acid synthesizer (custom built; Operon, Alameda, CA).
2. Sephadex G-25 columns (product no. 27-5325-01; Amersham Pharmacia, Piscataway, NJ).

3. 3X Saline sodium citrate (SSC).
4. 384-well plate (product no. Costar 6502; Corning, Corning, NY).
5. Arrayer (model no. PixSys 55000; Cartesian, Irvine, CA).
6. Arrayer pins (model no. ChipMaker 2; TeleChem, Sunnyvale, CA).

2.3. Postprocessing of Arrays

1. Humidity chamber (product no. H6644; Sigma).
2. Inverted heat block (Hotplate/Stirrer model no. PC-420; Corning).
3. Diamond scriber (product no. 52865-005; VWR Scientific).
4. Slide rack (product no. 121; Shandon Lipshaw). Each rack holds 30 slides.
5. Slide chamber (product no. 121; Shandon Lipshaw). Each chamber holds 350 mL.
6. Succinic anhydride (product no. 23,969-0; Aldrich, Milwaukee, WI).
7. 1-Methyl-2-pyrrolidinone (product no. 32,863-4; Aldrich).
8. 1 M Sodium borate, pH 8.0 (product no. 1330-43-4; VWR Scientific). Use boric acid and adjust pH with NaOH.
9. ddH$_2$O.
10. Two-liter beaker.
11. 95% ethanol.
12. 1X and 0.1X SSC.
13. Ultraviolet (UV) Crosslinker (product no. 400075; Stratagene, La Jolla, CA).
14. Blocking solution: 6 g of succinic anhydride in 325 mL of 1-methyl-2-pyrrolidinone, 15 mL of sodium borate.
15. Orbital shaker (model no. Innova 2000; New Brunswick Scientific).
16. Pyrex dishes.
17. Microwave.
18. Centrifuge (model no. Allegra 6R; Beckman Coulter).
19. Lamp (60 W).

2.4. Making cDNA for Cy Conjugation

1. Oligo-dT primer (oligo dT$_{18}$; Operon).
2. Superscript II RNase H- reverse transcriptase (200 U/mL; Life Technologies) and 5X first-strand buffer (Life Technologies).
3. 5-(3-Aminoallyl)-2'-deoxyuridine 5'-triphosphate Sodium Salt (AA-dUTP) (product no. A0410; Sigma).
4. dNTPs (dGTP, product no. D5038; dATP, product no. D4788; dCTP, product no. D4913; dTTP, product no. T9656, all from Sigma).
5. Reverse transcriptase mix: 6 µL of 5X first-strand buffer, 0.6 µL 25 mM AA-dUTP/dNTP (10 µL of 100 mM dATP, 10 µL of 100 mM dCTP, 10 µL of 100 mM dGTP, 6 µL 100 mM of dTTP, 4 µL of 100 mM AA-dUTP), 3 µL of dithiothreitol, 1.9 µL of Superscript II RNase H- reverse transcriptase, and 3 µL of diethylpyrocarbonate-treated H$_2$O.
6. 1 N NaOH.
7. 0.5 M EDTA.
8. 1 M Tris-HCl, pH 7.4 or 1 M HEPES, pH 7.0.

9. Microcon YM-30 (product no. 42409; Millipore, Bedford, MA).
10. ddH$_2$O.
11. MJ Research thermocycler (model no. PTC-200; MJ Research, Watertown, MA) and 0.2-mL RNase-free polymerase chain reaction (PCR) tubes.
12. Cy3 Mono-reactive dye pack (product no. PA23001; Amersham Pharmacia) and Cy5 Mono-reactive dye pack (product no. PA25001; Amersham). Cy aliquots: Resuspend each Cy dye in 72 μL of DMSO, aliquot 4.5 μL into 16 0.2-mL microcentrifuge tubes, dry in a Speed-Vac, and finally store the aliquots in the dark at 2–8°C.
13. Carbonate buffer (pH 9.6): 8 mL of 0.2 *M* sodium carbonate (Na$_2$CO$_3$), 17 mL of 0.2 *M* sodium bicarbonate (NaHCO$_3$), 25 mL of H$_2$O.
14. 4 *M* Hydroxylamine hydrochloride (product no. CSO996-2GM; VWR Scientific).
15. Qiagen PCR cleanup kit (product no. 28104; Qiagen, Valencia, CA). The kit is supplied with PB, PE, and EB buffers.
16. Microcentrifuge (Eppendorf Model 5415C; Brinkmann, Westbury, NY).

2.5. Hybridization, Washing, and Viewing of Arrays

1. 100°C heat block (Hotplate/Stirrer model no. PC-420; Corning).
2. Water bath (model no. 50; Precision Scientific, Winchester, VA).
3. Hybridization chamber (product no. 2551; Corning).
4. Cover slip, 22 × 22 mm (product no. 6776309; Shandon Lipshaw).
5. 3X SSC and 20X SSC.
6. 10% sodium dodecyl sulfate (SDS).
7. Fine-tip forceps (product no. 19040; Shandon Lipshaw).
8. Slide rack (product no. 121; Shandon Lipshaw).
9. Slide chamber (product no. 121; Shandon Lipshaw).
10. ddH$_2$O.
11. Wash solutions: solution 1 (1X SSC/0.03% SDS), solution 2 (0.2X SSC), and solution 3 (0.05X SSC). Formulations for the wash solutions are given in **Table 1**.
12. Array scanner (model ScanArray 3000, GSI Lumonics, Kanata, Ontario, Canada).

3. Methods
3.1. Preparation of Slides

Alternative to the following protocol, one can use commercially available amino-derivatized slides with 1,4 phenylene diisothiocyanate (PDC) attachment chemistry *(7)*. The PDC attachment chemistry works well when the printed volume of the oligonucleotide solution is 4–10 nL, with virtually no loss of probe oligonucleotide during the hybridization procedure. However, when the printed volume of the target oligonucleotide is below 1 nL, the sample dries too quickly and does not allow sufficient covalent bonding. Thus, the following procedure is recommended to avoid loss of the printed oligonucleotides during the hybridization and washing procedures.

Table 1
Preparation of Array Washing Solution

	Wash 1 (mL)	Wash 2 (mL)	Wash 3 (mL)
dH$_2$O	190	198	200
20X SSC	10	2	0.5
10% SDS	0.6	—^{a}	—^{a}
Final volume	200	200	200

^{a}No SDS was added.

1. Place the slides in the slide racks and then place the racks in the chambers.
2. Prepare the cleaning solution and pour into the chambers with the slides. Cover the chambers with glass lids. Mix on an orbital shaker for 2 h. Once the slides are clean, they should be exposed to air as little as possible (*see* **Note 1**).
3. Quickly transfer the racks to fresh chambers filled with ddH$_2$O. Rinse vigorously by plunging the racks up and down. Repeat rinses four times with fresh ddH$_2$O each time (*see* **Note 2**).
4. Prepare the polylysine solution and then transfer the slides into the polylysine solution and shake for 15 min to 1 h.
5. Transfer the rack to fresh chambers filled with ddH$_2$O. Plunge up and down five times to rinse.
6. Centrifuge the slides on microtiter plate carriers (place paper towels below the rack to absorb liquid) for 5 min at 500 rpm. Transfer the slide racks to empty chambers with covers for transport to a vacuum oven.
7. Dry the slide racks in a 45°C vacuum oven for 10 min (vacuum is optional).
8. Store the slides in a closed slide box (only plastic, without a rubber mat bottom).

3.2. Arraying of Oligonucleotides

Long oligonucleotides (70-mers, on average) complementary to the mRNA are designed to be near the 3' ends of genes and have hybridization temperatures very close to each other. In organisms, such as *S. cerevisae*, the designed oligonucleotide sequences can be BLASTed against the whole genome to ensure that there is no cross-hybridization owing to sequence relatedness or gene overlap. In humans, the designed sequences can be checked against known genes to eliminate cross-hybridization, or if one wishes to make a mammalian-general array, sequences can be chosen that cross-hybridize to humans, mice, and so on. For a schematic diagram of the following protocol, *see* **Fig. 1**.

1. Oligonucleotides were synthesized by Operon Technologies at the 1 or 0.2 µmol scale with or without a 5' C6 amino linker.
2. Desalt and purify the oligos on a Sephadex column, and then dry with a SpeedVac.

1. Synthesize oligos

↓

2. Desalt, Purify, and Dry

↓

3. Solubilize oligos and place in 384-well plate

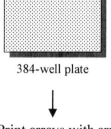

384-well plate

↓

4. Print arrays with arrayer

Fig. 1. Schematic diagram of arraying of oligonucleotides.

3. Solubilize and dilute to a final concentration of 40 μmol in 3X SSC and then pipet into 384-well plates (*see* **Note 3**).
4. The printing can be done with a variety of different arrayers; we use either a Cartesian or a custom-built version based on the Stanford design. We use TeleChem pins that deliver approx 0.5 nL and print the oligo probes in duplicate, triplicate, or quadruplicate. The replicates can be printed in different locations on the slides to minimize local depletion of the hybridization target in the hybridization solution. Printing should be done under humidity- and temperature-controlled environments; we use 60–70% humidity at 23°C.

3.3. Postprocessing of Arrays

1. Mark the boundaries of array on the back of a slide using a diamond scriber (*see* **Note 4**).
2. Fill the bottom of the humidity chamber with 100 mL of 1X SSC.
3. Prop the slide between two tip boxes, position a 60 W lamp overhead for better viewing, and etch the corners of the array.
4. Place the arrays facedown over 1X SSC and cover the chamber with a lid.
5. Rehydrate until the array spots glisten, approx 5–15 min. Allow the spots to swell slightly but not run into each other. Position the 60-W lamp approx 12 in. overhead for better viewing.
6. Snap-dry each array by placing (DNA side up) on a 70–80°C inverted heat block for 3 s.

7. UV crosslink the DNA to the glass with a Stratalinker set for 65 mJ. (Set the display to 650, which is 650×100 μJ.)

8. Place the arrays in the slide rack (*see* **Note 5**). Have the empty slide chamber ready on an orbital shaker.

9. Prepare the blocking solution: Dissolve 6 g of succinic anhydride in 325 mL of 1-methyl-2-pyrrolidinone. Rapid addition of the reagent is crucial.

10. Have three 350-mL slide chambers available (with metal tops) and a large round Pyrex dish with dH$_2$O ready in a microwave. At this time, prepare 15 mL of sodium borate.

11. Immediately after the last flake of the succinic anhydride dissolves (**step 9**), add the 15 mL of sodium borate. Immediately after the sodium borate solution mixes in, pour the solution into an empty slide chamber. Plunge the slide rack rapidly and evenly in the solution. Vigorously shake up and down for a few seconds, making sure the slides never leave the solution. Mix on an orbital shaker for 15–20 min. Meanwhile, heat water in the Pyrex dish (enough to cover the slide rack) to boiling.

12. Gently place the slide rack in 95°C water (just stopped boiling) for 2 min for double stranded products such as cDNAs or PCR products, and for oligonucleotide-based arrays gently place in 25°C water for 2 min or in 0.1X SSC for 5 min.

13. Plunge the slide rack five times in 95% ethanol.

14. Centrifuge the slides and rack for 5 min at 500 rpm. Load the slides quickly and evenly onto the carriers to avoid streaking.

15. Use the arrays immediately or store in a slide box.

3.4. Making cDNA for Cy Conjugation

The protocol utilizes monoreactive cyanine dyes to label cDNA after reverse transcription. Incorporation of a nucleotide containing an alkyl amino group (AA-dNTP) allows post-reverse transcription conjugation.

1. Reverse transcription: To anneal the primer, in a 0.2-mL RNase-free PCR tube, add 14.5 μL of total RNA (1 to 2 μg) and 1 μL of oligo-dT primer (65 μg/μL) or 1 μL of pd(N) (65 μg/μL). Incubate at 70°C for 10 min and cool on ice. Add 14.5 μL of reverse transcription mix and incubate at 42°C for 2 h.

2. Hydrolysis of RNA: Add 10 μL of 1 *N* NaOH and 10 μL of 0.5 *M* EDTA, pH 8.0. Incubate the sample for 15 min at 65°C. Neutralize the reaction by adding 25 μL of 1 *M* Tris-HCl, pH 7.4 or 25 μL of 1 *M* HEPES, pH 7.0.

3. Reaction cleanup: If Tris is used to neutralize the hydrolysis reaction, it must be removed prior to proceeding (*see* **Note 6**). A Microcon YM-30 can be used to remove the buffer. This is performed by placing 450 μL of H$_2$O in a Microcon YM-30 and next adding the reaction from **step 2**. Spin at 12,000 rpm for 12 min at room temperature and remove the flowthrough. Repeat wash twice with ddH$_2$O and then invert the Microcon in a new tube and elute by spinning 1 min. Eluent can be dried down in a speedvac and stored at –20°C.

4. Coupling of Cy dye: Resuspend cDNA in 4.5 μL of H$_2$O. Resuspend an aliquot of Cy dye in 4.5 μL of 0.1 *M* carbonate buffer. Mix cDNA and Cy dye and incubate at room temperature in the dark for 1 h.

5. Quench reaction: Nonreacted Cy dye can be quenched by adding primary amines. Add 4.5 μL of 4 *M* hydroxylamine hydrochloride and incubate for 15 min at room temperature in the dark.

6. Reaction cleanup II: Cy3 and Cy5 reactions can be combined and cleaned up together or separately. Removal of free dye can be done with either gel filtration via spin column or Pasteur pipet, or with various kits, such as the Qiagen PCR cleanup kit. The Qiagen PCR cleanup kit works well for this step (*see* **Note 7**). Protocol for Qiagen PCR cleanup kit: Add 35 μL of 100 m*M* NaOAc, pH 5.2, to each reaction. Combine the reactions in one tube. Add 500 μL of PB buffer. Apply the sample to a QIAquick column. Spin for 30–60 s, ~13,000 rpm (>10,000*g*). Discard flowthrough and add 750 μL of PE buffer. Spin for 30–60 s. Discard the flowthrough and spin for 1 min. Place the column in a new tube and add 50 μL of EB buffer or ddH$_2$O to the membrane. Spin for 1 min. The eluted volume is typically about 40 μL and can be reduced by a SpeedVac. Combined reactions should be in a volume of 18 μL (a SpeedVac can be used to control the volume or the volume can be dried and resuspended in H$_2$O).

7. Hybridization preparation: Add 3.6 μL of 20X SSC, 1.8 μL of polyA (10 mg/mL), and 0.54 μL of 10% SDS. Denature for 2 min at 95°C and apply to the microarray.

3.5. Hybridization, Washing, and Viewing of Arrays

1. Fluorescent DNA probes: Final probe volume should be 12–15 μL, at 4X SSC, containing competitor DNA, and so on, as required (e.g., polydA, CoT1 DNA for a human cDNA array).

2. Set up the array in the hybridization chamber. Place 10 μL of 3X SSC on the edge of the slide to provide humidity.

3. Add 0.3 μL of 10% SDS to the probe.

4. Boil the probe for 2 min. Set aside several cover slips for the next step.

5. Pipet the probe onto the array, avoiding bubbles. Using forceps, immediately place the cover slip over the array, avoiding bubbles.

6. Close the hybridization chamber and submerge in a water bath. Hybridizations should been done at 5–7°C below the lowest T_m of the oligonucleotide probes.

7. Hybridize for 4–24 h. With oligonucleotides the shorter time (4 h) works well.

8. Prepare wash solutions 1 (1X SSC/0.03% SDS), 2 (0.2X SSC), and 3 (0.05X SSC).

9. Pour the wash solutions in the slide chambers and place the slide racks in washes 1 and 2.

10. Disassemble the hybridization chamber and quickly submerge the array in wash 1 (*see* **Note 8**).

11. Let the array sit in wash 1 until the cover slip slides off. Gently plunge the rack up and down several times to wash the array; be sure not to scratch the array with the loose cover slip.

12. Manually transfer the array to the slide rack in wash 2 and rinse a second time.

13. Move the slide rack to wash 3 and rinse a third time. It is critical to remove all the SDS.

14. Centrifuge the slides on microtiter plate carriers (place paper towels below the rack to absorb liquid) for 5 min at 500 rpm at room temperature.

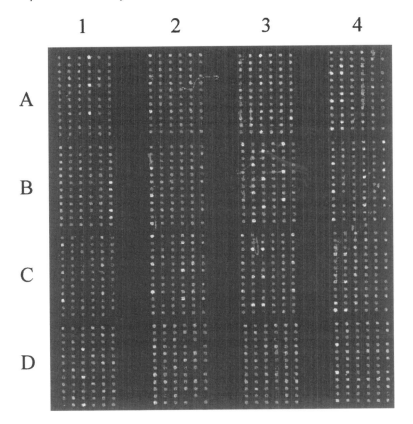

Fig. 2. Representative medium-density oligonucleotide array. Glass slide array gridded with (16) 6 × 12 array, 150 μm spots with a 500- and 350-mm pitch for the *x*- and *y*-axis, respectively. The array was hybridized with a standard quality control hybridization mixture of a random 9-mer oligonucleotide labeled with Cy3 that hybridizes to all spots with varying intensities. Areas 1A–4D represent separate sections of a spotter array containing 384 samples printed in triplicate. Dark gray spots (blue in color) indicates a relatively low level of hybridization (approx 3000 on a GSI scanner) and light gray spots (green and yellow in color) indicates a six- to sevenfold increase in signal (approx 20,000 on a GSI scanner).

15. Scan the array immediately (*see* **Note 9**). We use a ScanArray model 3000 from GSI Lumonics. **Figure 2** shows a representative medium-density oligonucleotide array.

4. Notes

1. One must be careful not to expose the slide to the air because minute dust particles will interfere with coating and printing.
2. It is critical to remove all traces of NaOH-ethanol.

3. Before printing arrays: Slides work best when "aged" for 3 to 4 wk before use. Check that the polylysine coating is not opaque. Test print, hybridize, and scan sample slides to determine slide batch quality.
4. It is important to mark boundaries on the back of the slides because arrays become invisible after postprocessing.
5. Be sure the rack is bent slightly inward in the middle; otherwise, the slides may run into each other while shaking.
6. The amine groups on the Tris-HCl can react with the monofunctional NHS-ester of the Cy dye.
7. It is important to keep in mind that the DNA-binding curve for silica, on which the Qiagen PCR cleanup kit is based, is favorable at a low pH but falls off precipitously around pH 8.0. Thus, it is essential that the pH of your reaction be below 7.5 by the time it hits the Qiagen membrane.
8. If the array is exposed to air while the cover slip starts to fall off, you may see high background fluorescent signal on the side of the array.
9. Be careful not to allow Cy5 bleaching from scanners with strong lasers. If the arrays are not scanned immediately, they should be stored in light-tight slide boxes until they are scanned.

References

1. Gillespie, D. and Spiegelman, S. A. (1965) A quantitative assay for DNA-RNA hybrids with DNA immobilized on a membrane. *J. Mol. Biol.* **12,** 829–842.
2. Ritossa, F., Malva, C., Boncinelli, E., Graziani, F., and Polito, L. (1971) The first steps of magnification of DNA complementary to ribosomal RNA in Drosophila melanogaster. *Proc. Nat. Acad. Sci. USA* **68,** 1580–1584.
3. Sinibaldi, R. M. and Cummings, M. R. (1981) Localization and characterization of rDNA in Drosophila tumiditarsus. *Chromosoma* **81,** 655–671.
4. Kafatos, F. C., Jones, C. W., and Estradiatis, A. (1979) Determination of nucleic acid sequence homologies and relative concentrations by a dot hybridization procedure. *Nucleic Acids Res.* **24,** 1541–1552.
5. Lockhart, D. J., Dong, H., Byrne, M. C., Follettie, M. T., Gallo, M. V., Chee, M. S., Mittmann, M., Wang, C., Kobayashi, M., Horton, H., and Brown, E. L. (1996) Expression monitoring by hybridization to high-density oligonucleotide arrays. *Nat. Biotechnol.* **14,** 1675–1680.
6. Schena, M, Shalon, D., Davis, R. W., and Brown, P. O. (1995) Quantitative monitoring of gene expression patterns with a complementary DNA microarray. *Science* **270,** 467–470.
7. Guo, Z., Guilfoyle, R. A., Thiel, A. J., Wang, R., and Smith, L. M. (1994) Direct fluorescence analysis of genetic polymorphisms by hybridization with oligonucleotide arrays on glass supports. *Nucleic Acids Res.* **22,** 5456–5465.

15

Use of Bioinformatics in Arrays

Peter Kalocsai and Soheil Shams

1. Introduction

Previous chapters have discussed in detail how to prepare, print and hybridize DNA arrays on various surfaces. Once hybridization is completed, the next step is to scan in the glass, gel, or plastic slides with a specialized scanner to obtain digital images of the results of the experiment. The DNA expression levels are then quantified with the help of image-analysis software. After the image processing and analysis step is completed, we end up with a large number of quantified gene expression values. The data typically represent hundreds or thousands, in certain cases tens of thousands, of gene expressions across multiple experiments. To make sense of this much information requires the use various of visualization and statistical analysis techniques. One of the most typical microarray data analysis goal is to find statistically significantly up- or downregulated genes; in other words outliers or "interestingly" behaving genes in the data. Other possible goals could be to find functional groupings of genes by discovering similarity or dissimilarity among gene-expression profiles, or predicting the biochemical and physiological pathways of previously uncharacterized genes.

Before any of that analysis could take place, one has to address the issues of normalization and possible transformation of the data so that, as much as possible, the quantified values would only represent true differences among gene expressions. On the other hand, finding a transformation that would make reading the data more understandable, or would change the distribution of the data in a way that it would be more suitable for certain statistical analysis, could also be an important goal. In **Subheading 2.**, we discuss all these issues followed by a description of tools and algorithms implementing various

From: *Methods in Molecular Biology, vol. 170: DNA Arrays: Methods and Protocols*
Edited by: J. B. Rampal © Humana Press Inc., Totowa, NJ

multivariate techniques that could help solve the data analysis and visualization needs of the DNA array research community in **Subheading 3.**

2. Data Normalization and Transformation
2.1. Examination of the Data

Before even starting to analyze the DNA array data, what one has to be aware of is that no matter how powerful statistical methods are used the analysis is still crucially dependent on the "cleanness" and distributional properties of the data. Essentially, the two questions to ask before starting any analysis are:

1. Does the variation in the data represent true variation in expression values or it is contaminated by differences in expression due to experimental variability?
2. Is the data "well-behaving" in terms of meeting the underlying assumptions of the statistical analysis techniques that are applied to it?

It is easy to appreciate the importance of the first point. The significance of the second one comes from the fact that most multivariate analysis techniques are based on underlying assumptions such as normality and homoscedasticity. If these assumptions are not met at least approximately, then the whole statistical analysis could be distorted and statistical tests might be invalid. Fortunately, there are a variety of statistical techniques available to help us answer "Yes" to the above two questions respectively:

1. Normalization (standardization).
2. Transformation.

Normalization, and as a special form of normalization standardization, can help us separate true variation from differences due to experimental variability. This step might be necessary, as it is quite possible that due to the complexity of creating, hybridizing, scanning, and quantifying microarrays variation originating from the experimental process contaminates the data. During a typical microarray experiment, many different variables and parameters can possibly change, hence differentially affect, the measured expression levels. Among these are slide quality, pin quality, amount of DNA spotted, accuracy of arraying device, dye characteristics, scanner quality, and quantification software characteristics, to name a few. The various methods of normalization aim at removing, or at least minimizing, expression differences due to variability in any of these types of conditions.

As will be discussed later, the various transformation methods all aim at changing the variance and distribution properties of the data in such a way so that it would be closer in meeting the underlying assumptions of the statistical techniques applied to it in the analysis phase. The most common requirements of statistical techniques are for the data to have homologous variance (homoscedasticity) and normal distribution (normality).

Several popular ways of normalizing and transforming microarray data are discussed as follows.

2.2. Normalization of the Data

One of the most popular ways to control for spotted DNA quantity and other slide characteristics is to do a type of local normalization by using two channels (red and green) in the experiment. For example, a cy5 (red)-labeled probe could be used as control prepared from a mixture of cDNA samples or from normal cDNA. Then a cy3 (green)-labeled experimental probe could be prepared from cDNA extracted from a tumor tissue. The normalized expression values for every gene would then be calculated as the ratio of experimental and control expression. This method can obviously eliminate a great portion of experimental variation by providing every spot (gene) in the experiment with its own control. Developing on these ideas, three-channel experiments are underway where one channel serves as control for the other two. In this case, the expression values of both experimental channels would be divided by the same control value.

In addition to the foregoing-described local normalization, global methods are also available in the form of "control" spots on the slide. Based on a set of these control spots, it becomes possible to control for global variation in overall slide quality or scanning differences.

These procedures describe some physical measures in terms of spotting characteristics that one can take to normalize the microarray data. However, even after the most careful two or more channel spotting with the use of control spots it is still possible that undesired experimental variation contaminates the expression data. On the other hand, it is also possible that all or some of these physical normalization techniques are missing from the experiment in which case it is even more important to find other ways of normalization. Fortunately, for both of these scenarios additional statistics based normalization methods are available to further clean up the data.

For example, the situation can happen when gene expression values in one experiment are consistently and significantly different from another experiment for the same set of genes due to quality differences between the slides or the printing or scanning process or possibly due to some other factor (*see* **Fig. 1**).

It might be very misleading to compare the expression values of the two files plotted in **Fig. 1** without any normalization. When subtracting the mean value and dividing by the standard deviation in both experiments we can make a much more realistic comparison. The same data are plotted in **Fig. 2** after normalization. Notice that most of the gene expressions now lay on the identity line as it would be expected for two experiments on the same set of genes.

We might mention that statistics based normalization or standardization can be accomplished in many different ways involving either the mean, median,

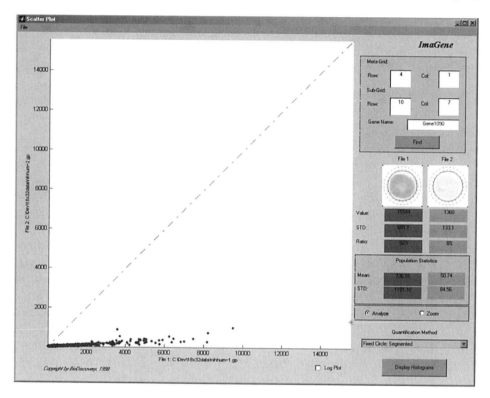

Fig. 1. Scatterplot of two experiments where the signal values are significantly stronger in one file then in the other. File 1 is the red value and File 2 the green. Reprinted with permission from **ref. 8**.

mode, or possibly some other statistics of the data. As a general rule, the choice of statistics should depend on the distributional properties of the data.

2.3. Transformation of the Data

Although there are many different data transforms available, the most frequently used procedure in the microarray literature is to take the logarithm of the quantified expression values *(1,2)*. An often-cited reason for applying such a transform is to equalize variability in the typically wildly varying raw expression scores. If the expression value was calculated as a ratio of experimental over control conditions then an additional effect of the log-transform will be to equate up- and downregulation by the same amount in absolute value scores ($\log_{10}2 = 0.3$ and $\log_{10}0.5 = -0.3$). Another important side effect of the log transform is bringing the distribution of the data closer to normal. Having reasonable grounds of meeting the normality and homoscedasticity assump-

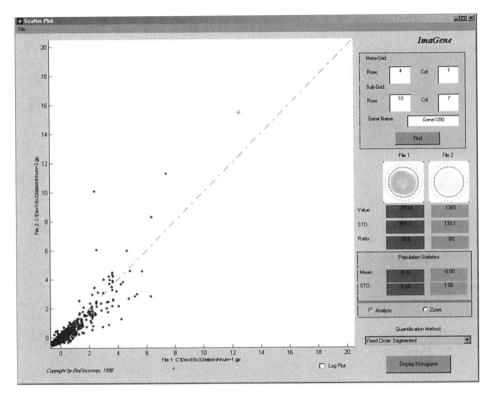

Fig. 2. The same data as in **Fig. 1** after normalization. File 1 is the red value and File 2 the green. Reprinted with permission from **ref. 8**.

tions after the log transform it is now much better justified to use a variety of parametric statistical analysis methods on the data.

As shown in **Fig. 3** (left panel), without the log transform we get a very peaked distribution of the data with a very long positive tail. This distribution is very far from normal, which violates the assumptions made by many standard parametric statistical analysis methods. The distribution of the log-transformed data in the right panel is visibly much closer to that of the normal distribution, which could also be verified by a simple normality test. Certainly, the number of transforms that could be experimented with is almost limitless, but the most frequently applied once are $\log(x)$, $\sqrt{x}$, $1/x$, and arcsin $\sqrt{x}$. In fact, the choice of transformation should be dependent on which one brings the data closest to the requirements of homogeneity of variance and normal distribution *(3)*. It turns out that for microarray expression data the usually applied log transform provides the best solution. On the other hand when nonparametric analysis techniques are used, which have no restrictions on the distributional

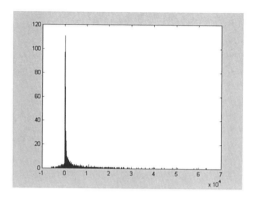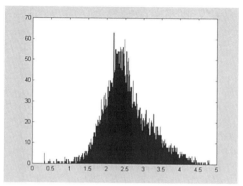

Fig. 3. Histograms of gene expression data involving 600 genes across 21 experiments. The x-axis indicates the expression values and the y-axis shows the number of genes with a particular gene expression level. The left and right panels show the data before and after the log transform, respectively. Reprinted with permission from **ref. 8**.

properties of the data, the transformation step might not be necessary at all. This flexibility of nonparametric methods comes at a price of usually requiring more data and, in many cases, providing less accuracy then parametric techniques.

3. Data Analysis

In **Subheading 2.**, we gave an overview of the most important data preprocessing steps that one might consider before starting analyzing the data. In what follows we describe, with increasing complexity, the analysis techniques that appear to be the most useful for DNA array data.

3.1. Scatterplot

Probably the simplest analysis tool for microarray data visualization is the scatterplot, as introduced earlier. In a scatterplot, each point represents the expression value of a gene in two experiments, one plotted on the x-axis and the other one on the y (*see* **Fig. 2**). In such a plot, genes with equal expression values would line up on the identity line (diagonal) with higher expression values further away from the origin. Points below the diagonal represent genes with higher expression in the experiment plotted on the x-axis. Similarly, points above the diagonal represent genes with higher expression values in the experiment plotted on the y-axis. The further the point away is from the identity line, the larger is the difference between its expression in one experiment compared with the other.

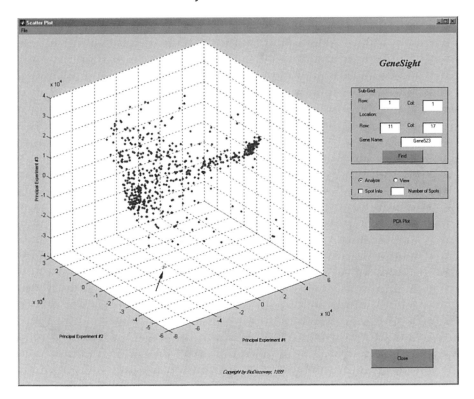

Fig. 4. Principal component analysis on 600 genes across 20 experiments. Gene scores are plotted on the first three principal components. The gene pointed to by the arrow shows a possible outlier.

3.2. Principal Component Analysis

It is easy to see how the scatterplot is an ideal tool for comparing the expression profile of genes in two experiments. Even three experiments could be plotted and compared in a 3-dimensional scatterplot. What can we do though when more than 3 experiments are to be analyzed and compared with each other? In case of 20 experiments, for example, we cannot draw a 20-dimensional plot. Fortunately, there are techniques available in statistics for dimensionality reduction, such as principal component analysis, which are able to compress the data into two or three dimensions (that we can plot) while preserving most or all the variance of the original dataset. **Figure 4** is in fact a 3-dimensional plot of 600 genes in 21 experiments indicating the scores of all 600 genes on the first three principal components. Of course, lower-ranked principal components could also be plotted, three or less at a time, with the understanding that they account for less and less of the overall variance in the data.

This multivariate technique is frequently used to provide a compact representation of large amounts of data by finding the axes (principal components) on which the data varies the most. In principal component analysis, the coefficients for the variables are chosen such that the first component explains the maximal amount of variance in the data. The second principal component is perpendicular to the first one and explains maximum of the residual variance. The third component is perpendicular to the first two and explains maximum of the still remaining variance. This process is continued until all the variance in the data is explained. The linear combination of gene expression levels on the first three principal components could easily be visualized in a three-dimensional plot (*see* **Fig. 4**). This method, just like the scatterplot earlier, provides an easy way of finding outliers, in the data; genes that behave differently than most of the genes across a set of experiments. With a transpose of the data matrix, the experiments could also be plotted to find out possible groupings and/or outliers of experiments. Recent findings show that this method should be able to even detect moderate-sized alterations in gene expression *(2)*. In general, principal component analysis provides a rather practical approach to data reduction, visualization, and identification of unusually behaving outlier genes and/or experiments.

3.3. Parallel Coordinate Planes

Two- and three-dimensional scatterplots and principal component analysis plots are ideal for detecting significantly up- or downregulated genes across a set of experiments. These methods do not provide an easy way of visualizing progression of gene expression over several experiments, however. These types of questions usually come up in time series experiments where gene expression is measured every two hours, for instance. The important question in this case is how gene expression progresses over the duration of the entire experiment. The parallel coordinate planes plotting technique is best suited to answer these types of questions. With this method experiments are ordered on the *x*-axis and expression values plotted on the *y*-axis. All genes in a given experiment are plotted at the same location on the *x*-axis; only their *y* location varies. Another experiment is plotted at another *x* location in the plane. Typically, the progression of time would be mapped into the *x*-axis by having higher *x* values for experiments done at a later time or vice versa. By connecting the expression values for the same genes in the different experiments one can obtain a very intuitive way of depicting the progression of gene expression (*see* **Fig. 5**).

Showing changes in expression pattern during the cell's life cycle can readily be visualized with parallel coordinate planes. Not only does this make this type of display very easy to follow the changes in expression level over time, but it could well be applied to any other type of data as well, such as comparison of

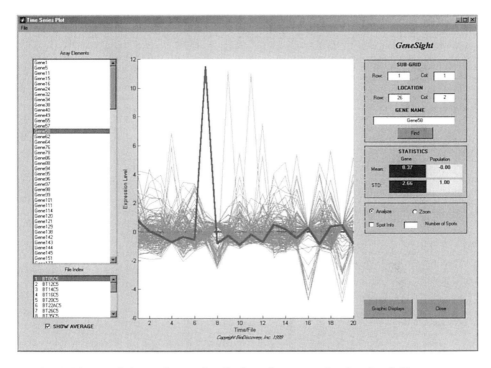

Fig. 5. The parallel coordinate plot displays the expression levels of all genes across all experiments/files in the analysis. On the *x*-axis the experiments or experimental files are plotted. The *y*-axis shows the expression level of all genes across all experiments.

expression level in different tissue types, for example. Due to the easy detection of unusual expression patterns, this type of plot can also be used well for outlier detection.

In addition, by applying different curve-fitting techniques any expression pattern over time could be searched for in the data. By adjusting the required closeness of fit the number of chosen expression patterns could also be controlled.

3.4. Cluster Analysis

Another frequently asked question related to microarrays is finding groups of genes with similar expression profiles across a number of experiments. The most often-used multivariate technique to find these groups is cluster analysis. Essentially, this method accomplishes the sorting of the data with grouping (clustering) genes with similar expression patterns closer to each other. This technique can help establish functional groupings of genes or predicting the biochemical and physiological pathways of previously uncharacterized genes.

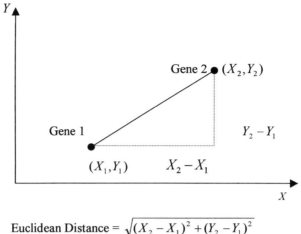

$$\text{Euclidean Distance} = \sqrt{(X_2 - X_1)^2 + (Y_2 - Y_1)^2}$$

Fig. 6. Illustration of calculating the Euclidean distance between two genes in two experiments.

The clustering method that is most frequently used in the literature for finding groups in microarray data is hierarchical clustering *(1)*. This method typically operates on a similarity or distance measure of the data such as correlation, Euclidean, squared Euclidean, or city-block (Manhattan) distance. To demonstrate the meaning of distance between two gene expressions values in two experiments, the calculation of the Euclidean distance is shown below in **Fig. 6**. The expression values for the two experiments are plotted on the *X*- and *Y*-axes, respectively. The expression values for a gene in the two experiments were and and when plotted they produce the point "gene 1" in the *X–Y* plane. The point "gene 2" is produced similarly. From the Pythagorean theorem, the Euclidean distance between points gene 1 and gene 2 can easily be calculated.

This example was two-dimensional, as there were only two experiments, but it can easily be extended to *N* dimensions, where *N* is the number of experiments in the study. Calculating the correlation matrix of the data is even less extensive computationally than the Euclidean distance, and in many situations it already produced quite sufficient results *(1,4)*.

In addition to calculating the correlation or distance matrix, in most cases a linkage rule also has to be specified to indicate how distance should be calculated between groups and when groups are supposed to be joined together. The most popular linkage rules are single, complete, average linkage, or the centroid method. For example, in the average-linkage method, the distance between two clusters is calculated as the average distance between all pairs of objects in the two different clusters. As a result of the grouping process, a tree

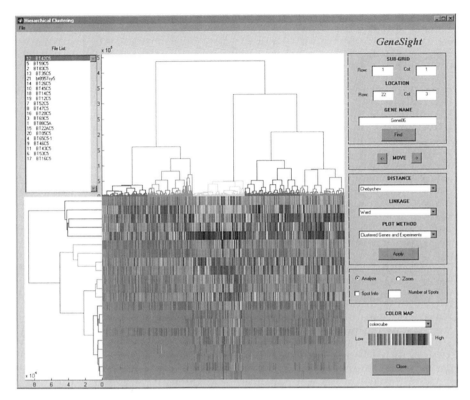

Fig. 7. Color-coded gene expression values for 600 genes (horizontally) in 21 experiments (vertically). Simultaneous clustering of genes and experiments is visualized by the top and left-side dendrograms, respectively.

of connectivity of observations emerges that can easily be visualized as dendrograms. For gene expression data, not only the grouping of genes but the grouping of experiments might also be important. When both are considered, it becomes easy to simultaneously search for patterns in gene expression profiles and across many different experimental conditions (*see* **Fig. 7**).

Every colored block in the middle panel of **Fig. 7** represents the expression value of a gene in an experiment. The 600 genes are plotted horizontally and the 21 experiments are plotted vertically. The color code is located in the lower-right corner. The dendrogram for genes is located just above the color-coded expression values with one arm connected to every gene in the study. As shown in **Fig. 7**, there are three major groups of genes in the study: The group on the left represents genes that have about medium-level expression in most experiments, but there also some experiments in the lower part with low expression. The middle, smaller, group represents genes that maintained medium-level expression across all experiments with very high expressions in some of the experiments in the top part. The third group shows genes with low, medium, and high expression levels in certain experiments. These three major groups could then be subdivided further into smaller groups as shown in **Fig. 7**. The dendrogram for experiments is on the left showing the grouping of the 21 experiments in the study. The experiments with high overall expression levels are clustered together on the top. Experiments with medium-level expression are in the middle, whereas experiments with low-level expressions are grouped together in the bottom of the color-coded figure.

Although currently hierarchical clustering is an often employed way of finding groupings in the data, other nonhierarchical (*k* means) methods are likely to gain more popularity in the future with the rapidly growing amounts of data and the ever increasing average experiment size. A common characteristic of nonhierarchical approaches is to provide sufficient clustering without having to create the full distance or similarity matrix, but with minimizing the number of scans of the whole data set.

Cluster analysis is currently by far the most frequently used multivariate technique to analyze gene expression data. We have to emphasize, however, that it is also the simplest such method. Cluster analysis is typically employed when there is no *a priori* knowledge about the data available. We are at the very beginning of understanding the gene interaction network of even some of the simplest genomes, but it would certainly be misleading to say that nothing is known about the functionality of genes in certain genomes. For example, the Munich Information Center for Protein Sequences Yeast Genome Database classifies genes belonging into functional classes such as: tricarboxylic-acid pathway, respiration chain complexes, cytoplasmic ribosomes, and many others *(5)*. Many of these functional categories represent genes, which, on biological grounds, are expected to have similar expression profiles across a set of experiments *(1,4)*. One could, of course, apply the previously described clustering scheme to group genes with similar expression profiles and from the known genes in each group conclude which group represents which biological functionality. With such a procedure, one might find that the clustering actually came up with groups that biologically make sense, but the opposite is equally possible. Depending on the chosen algorithm, some of its parameters or just due to characteristics of the data it is also possible that the found clustering has no biological significance at all.

4. Conclusions

In this chapter, we gave an overview of the most popular data analysis and visualization techniques that are used with microarray expression experiments. In our discussion, we started out with the simplest tools such as the scatterplot, principal component analysis, and showing the data in parallel coordinate planes with gradually working our way toward the more complex analysis methods such as the various forms of clustering methods. The discussion should give the reader an overview of the currently used most popular analysis techniques with a peek about what to expect in the near future.

On a related note, no one would argue that even with the best possible algorithms the result of an analysis is crucially dependent on the quality of the data. Because of that importance, a whole section is committed to the various ways how one could improve data quality. That is, the numerous normalization and data transformation methods are discussed that have already proved useful for microarray data or at least have a chance of showing applicability in future analyses.

Currently, cluster analysis is the most popular multivariate technique that is used to find structure in microarray data. As pointed out earlier it is not without limitations, but as probably the simplest possible multivariate technique it quickly gained popularity. The authors predict that, as the field matures, we are likely to see a shift in the direction of more sophisticated classification methods appearing in the literature. Even though classification is certainly a more direct way of finding structure in the data than clustering, it still lacks the complexity that is required to capture all the connectivity and interdependence among genes in a genome.

Keep in mind that probably the ultimate goal of analyzing microarray data is, at some point, to discover how genes are related and affect each other, and dependent on one another. Accordingly, the modeling device describing this interdependency has to have a matching level of complexity. There are not too many modeling tools out there that would fit these requirements. Some of the possible candidates are multilayer neural networks, system of partial differential equations and structural equation modeling. At this point, it would be too early and also hard to tell which one or several of these and other methods will turn out to be the most applicable modeling tool(s), but with the rapidly accumulating expression data these techniques are bound to appear in the relatively near future. Some early examples of applying neural networks to explain gene data *(6)* and mapping out the connectivity pattern of smaller regulatory networks *(7)* are already available.

References

1. Eisen, M. B., Spellman, P. T., Brown, P. O., and Botstein D. (1998) Cluster analysis and display of genome-wide expression patterns. *Proc. Natl. Acad. Sci.* **95,** 14,863–14,868.
2. Hilsenbeck, S. G., Friedrichs, W. E., Schiff, R., O'Connell, P., Hansen, R. K., Osborne, C. K., and Fuqua, S. A. W. (1999) Statistical analysis of array expres-

sion data as applied to the problem of tamoxifen resistance. *J. Natl. Cancer Inst.*
91, 453–459.

3. Ferguson, G. A. and Takane Y. (1998) *Statistical Analysis in Psychology and Education*, McGraw-Hill, New York.

4. Brown, P. O. and Botstein, D. (1999) Exploring the new world of the genome with DNA microarrays. *Nat. Genet. Suppl.* **21,** 33–37.

5. Munich Yeast Genome Database (1999) Munich information center for protein sequences yeast genome database. http://www.mips.biochem.mpg.de/proj/yeast.

6. Yuh, C., Bolouri, H., and Davidson, E. H. (1998) Genomic cis-regulatory logic: experimental and computational analysis of a sea urchin gene. *Science* **279,** 1896–1902.

7. Tamayo, P., Slonim, D., Mesirov, J., Zhu Q., Kitareewan, S., Dmitrovsky, E., Lander, E. S., and Golub, T. R. (1999) Interpreting patterns of gene expression with self-organizing maps: Methods and application to hematopoietic differentiation. *Proc. Natl. Acad. Sci.* **96,** 2907–2912.

8. Zhou, Y., Kalocsai, P., Chen, J., and Shams, S. (2000) Information processing issues and solutions associated with microarray technology, in *Microarray Biochip Technology, BioTechniques Books Publication*, (Schena, M., ed.), Eaton Publishing, Natick, MA.

16

Confocal Scanning of Genetic Microarrays

Arthur E. Dixon and Savvas Damaskinos

1. Introduction

Confocal scanning laser microscopes are best known for making very high-resolution images of small three-dimensional (3D) specimens by recording a series of two-dimensional confocal slices of the specimen at different focus positions. Additionally, confocal microscopes are excellent for measuring weak fluorescence from areas adjacent to areas of very strong fluorescence, because the confocal pinhole rejects light from areas outside the focus spot. The resolution of a confocal microscope is also better than that of a nonconfocal instrument, an important property when submicron resolution is required. At first glance, none of these properties seems important for imaging genetic microarrays, whose features range from 200 to 10 μ in size, and may cover the entire surface of a microscope slide in a two-dimensional array. However, we will find that the array of genetic material, when deposited on a weakly fluorescent substrate (like a glass slide) or in contact with an aqueous buffer solution acts like a 3D specimen as far as imaging is concerned. Confocal imaging helps to reject much of the background signal from the glass slide or the aqueous solution, and it is also useful for rejecting fluorescence from areas that surround the focal spot in the focal plane.

1.1. Review of Confocal Microscopy

An excellent source for general information on confocal microscopy is the Handbook of Biological Confocal Microscopy (1). **Figure 1** shows a basic confocal microscope. Light from a point source (often a pinhole illuminated by a focused laser beam) passes through a beam splitter, expands to fill a microscope objective, and is focused to a tiny volume (the focal point) inside the specimen (solid lines in **Fig. 1**). Light reflected (or emitted) from that point in

From: *Methods in Molecular Biology, vol. 170: DNA Arrays: Methods and Protocols*
Edited by: J. B. Rampal © Humana Press Inc., Totowa, NJ

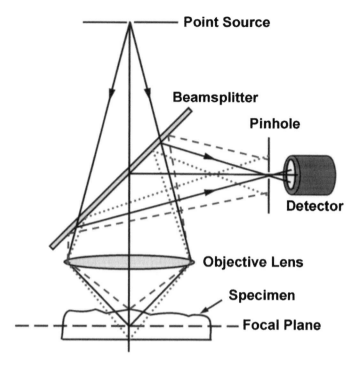

Fig. 1. Simple confocal microscope.

the specimen is collected by the objective lens, and is partially reflected to the right to pass through a pinhole to reach the detector. At the same time, light is reflected from parts of the specimen above the focal point (dashed lines in **Fig. 1**). This light is also collected by the objective lens and is partially reflected to the right, converging toward a focus behind the pinhole. Most of the light reflected from this point in the specimen runs into the metal surrounding the pinhole and is not detected. Similarly, light reflected from parts of the specimen below the focal point (dotted lines in **Fig. 1**) converges toward a focus in front of the pinhole, and then expands to hit the metal area surrounding the pinhole. Again, this light is blocked from reaching the detector. Thus, the pinhole blocks light from above or below the focal point, so the detector output is proportional to the amount of light reflected back from only the parts of the specimen at the focal point. Images of the source pinhole and the detector pinhole are formed by the objective lens at the focal point (the source pinhole, detector pinhole, and the focal point are "confocal" with each other).

An image is collected by moving the specimen under the fixed laser beam in a raster scan (a scanning-stage microscope), by moving the beam using mirror scanners (a scanning-beam system), by moving the objective lens (a scanning-

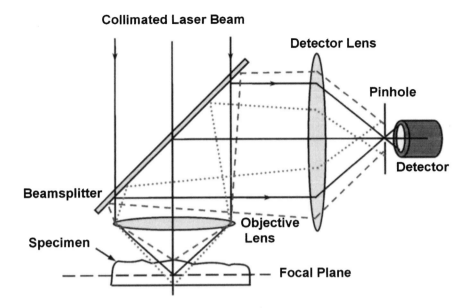

Fig. 2. Infinity-corrected confocal microscope.

head system), or by moving the beam in one direction while moving the specimen in the perpendicular direction. Confocal images consist of sharp and empty areas; nonconfocal images consist of sharp and blurry areas. In a confocal microscope, light from out-of-focus parts of the specimen is rejected by the confocal pinhole, and it is this absence of out-of-focus information that allows 3D images to be formed using a series of confocal slices.

For many applications, an infinity-corrected confocal microscope is more useful than the simple microscope in **Fig. 1**. In an infinity-corrected microscope (*see* **Fig. 2**), a parallel beam from a laser or other light source is focused by an infinity-corrected microscope objective onto a specimen at the focal plane. Light reflected from the specimen is collected by the microscope objective and partially reflected by the beam splitter toward a detector lens that focuses the beam to pass through the detector pinhole to reach the detector. As before, light reflected (or emitted) from above or below the focal plane is rejected by the pinhole, and the microscope is confocal.

An infinity-corrected arrangement has many advantages. The length of the detection arm can be changed without any effect on the focus position at the detector pinhole. Filters required for rejection of the laser wavelength in fluorescence imaging may be added to the detection arm without affecting the focus at the detector pinhole, because they are being inserted into a parallel beam. Finally, it is easier to incorporate scan mirrors into the system.

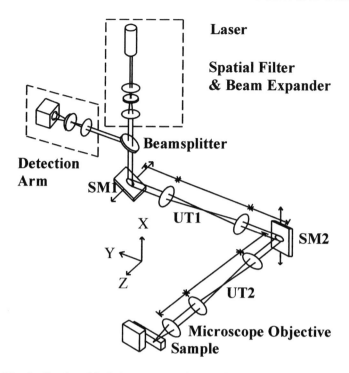

Fig. 3. Confocal infinity-corrected scanning-beam microscope.

Figure 3 presents an infinity-corrected scanning beam confocal microscope. The laser beam passes through a lens focused on a pinhole, and is collected by a second lens one focal length from the pinhole, resulting in a parallel beam. The focal length of the second lens is chosen to produce a beam diameter that will fill the entrance pupil of the microscope objective. This combination of a lens, pinhole, and second lens is called a *spatial filter*. The expanded laser beam passes through a beam splitter and is reflected from a mirror scanning about the z-axis. A unitary telescope brings this beam back to the center of the second scanning mirror, which scans about the x-axis. After reflection from the second scanning mirror, a second unitary telescope brings the scanning beam back to the axis of the microscope at the entrance pupil of a microscope objective. The beam is focused to a diffraction-limited spot at the sample, which is mounted on a stage that moves in the z-direction for focusing. Light reflected back from the sample (or fluorescence emitted from the sample) is collected by the microscope objective, descanned by the scanners, and partially reflected by the beamsplitter to enter the detection arm. In the detection arm, a lens focuses the parallel beam onto a pinhole, and its intensity is measured by a detector placed behind the pinhole. For fluorescence imaging, the beamsplitter usually used is a dichroic beamsplitter, chosen to reflect light of wavelength longer than the laser wavelength, and additional filters may also be used in the detection arm to further increase rejection of the laser light.

1.2. Genetic Microarray Imaging

Two types of microarrays are in widespread use. Glass slide microarrays are composed of a square or rectangular array of round spots of DNA material that have been deposited on glass microscope slides by a robotic spotter. These spots are usually between 50 and 200 µ in diameter. Two or more fluorophores are often used, with Cy3 and Cy5 the most popular choice. After hybridization, the spots are usually allowed to dry before reading. Because of the large size of the spots of DNA material, a focus spot size of 10 µ is small enough to provide good resolution for imaging the array. A dynamic range of 16 bits is required for detection. Good confocality rejects residual fluorescence in the glass microscope slide as well as light from surrounding fluorescent areas on the slide.

Another commonly used microarray is the Affymetrix GeneChip® (*see* **Note 1**). These chips are smaller than glass slide microarrays (often 1.2×1.2 cm) but contain a very large number of square probes (as small as 10×10 µ). These small probes require a focus spot size of approx 2 µ for adequate imaging. Another difference is that the probes are mounted on the inside surface of a thin glass plate that is permanently attached to a small hybridization chamber. After hybridization, the chamber is filled with a buffer solution, so the chip is wet when read. This means that the chip should be read in the vertical plane so that no bubbles form on the surface. In addition, the scanner should have good confocality in order to reject any residual fluorescence in the buffer solution, and there should be no rapid motion of the sample during scanning.

Genetic microarrays are too large to be imaged in a conventional confocal scanning beam laser microscope. Microscope objectives are available that have a field of view of more than 1 cm, but they have very low numerical apertures and are not suitable for imaging weak fluorescent samples. Suitable imaging systems include those based on scanning-stage and scanning-head confocal microscopes and the confocal scanning beam MACROscope® *(2,3)* (*see* **Note 2**). The DNAscope™ (*see* **Note 3**) from *GeneFocus*™ (*see* **Note 3**) described here and illustrated in **Fig. 4** is based on the MACROscope® technology and is suitable for scanning both glass slide microarrays and Affymetrix GeneChips.

The DNAscope works as follows. The laser beam is focused on a pinhole by a lens, and the expanding beam exiting the pinhole is collimated by a second lens into a parallel beam approx 2 cm in diameter, to match the size of the entrance pupil of the laser scan lens. The beam passes through a beam splitter and is reflected from a pair of computer-controlled scan mirrors positioned on either side of the entrance pupil of the laser scan lens. The laser scan lens

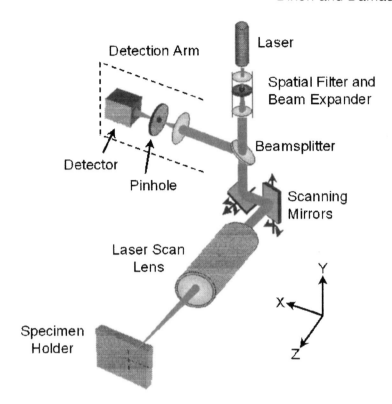

Fig. 4. DNAscope with telecentric *f**theta laser scan lens (*see* **ref. 3**).

focuses the incoming beam to a 10-µ spot size on the microarray (2-µ spot size in the GeneChip version). Fluorescence from the specimen is collected by the laser scan lens and descanned by the scanners, and the stationary beam traveling back toward the laser is reflected by the dichroic beam splitter (chosen to reflect light whose wavelength is longer than the laser wavelength) into the detection arm. Here the beam passes through a laser rejection filter (not shown), and is focused by a lens onto the detector pinhole, which is confocal with the focal spot at the specimen, and the detector measures the beam intensity. An image of the microarray is generated by scanning the focused laser spot across the microarray in a raster scan, while the output of the detector is digitized at a constant rate (*see* **Note 4**).

2. Genetic Microarray Image and Data Analysis

Figure 5 presents an image of a glass slide microarray recorded with the DNAscope. This yeast array has been tagged with both Cy3 and Cy5 fluorophores; the Cy3 image is shown. The array was deposited on a glass

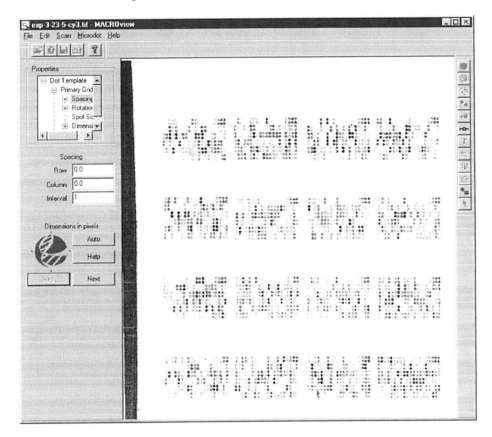

Fig. 5. Fluorescence image of a glass slide microarray.

microscope slide using a robotic spotter with a 16-pin head. This resulted in 3840 spots of 120 µ diameter in 16 subarrays inside a 2 × 2 cm area on the glass slide. This 2048 × 2048 pixel image shows the raw data before any processing has begun, with 16-bit dynamic range, and with the contrast reversed for better visibility of the spots. Imaging time was 3 min.

MACROview (*see* **Note 3**) software was used to analyze the data. One subarray is selected by placing a box around it using the mouse. An automatic spot detector then calculates horizontal and vertical spacing, horizontal and vertical tilt, and spot size. The results of these calculations are shown in a dialogue box on the edge of the computer screen. A grid is selected by inputting the number of rows and columns in the subarray and is anchored by clicking on the top left spot. This grid is then displayed, and reference circles slightly larger than the spot size are automatically located on the grid. At this stage, all the spots in the subarray may not be centered inside the circles.

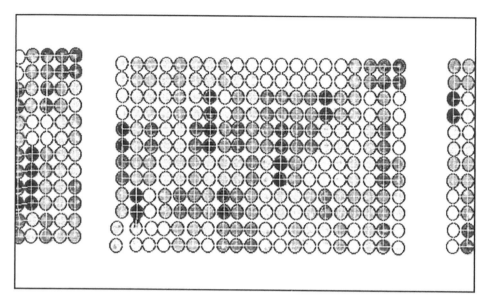

Fig. 6. Subarray showing grid and reference circle placement by the automated circle placement algorithm.

The next step is to copy the grid (still not perfectly aligned with all the spots) to all the other subarrays by clicking on the top left spot in each subarray. An automatic circle placement algorithm then adjusts the position of the reference circles inside the selected subarray, and finally across the entire image. **Figure 6** shows the result at this stage. **Figure 6** has been zoomed in to show one subarray and parts of the subarrays on either side. Both the grid and circles are shown, with the fluorescent probe spots inside the automatically placed circles. This subarray was chosen because it contains a pair of misaligned probe spots at the bottom left, which is a challenge for the automatic circle placement algorithm, and the subarray on the left is not aligned with that in the center of the picture. Although the algorithm has worked well here, it may not provide perfect placement when adjacent spots overlap, so manual placement tools are provided. A grid of squares can be used instead of circles, as shown in **Fig. 7**. This feature was developed for use with Affymetrix GeneChips, but also works well for glass slide microarrays.

The final step comprises integration of the fluorescence intensity inside each reference circle or square, calculation of local background fluorescence by integrating over four small areas between the probe spots, subtraction of the local background from each spot, and calculation of the relative intensity of each spot on the array.

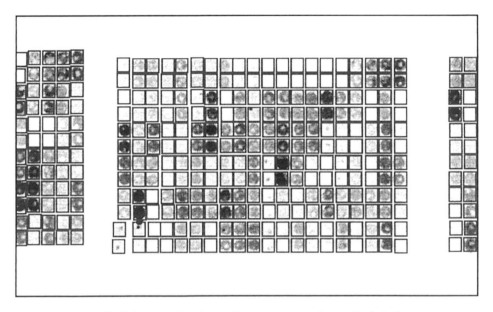

Fig. 7. Subarray showing reference squares instead of circles.

Measurement and subtraction of local fluorescence background is a critical step, because it is at this stage that the zero signal level is set. Some scanners (including the DNAscope) allow the operator to specify a detector offset that removes the dark current signal from the photomultiplier tube. This offset can also be used to remove a global background signal, e.g., a fluorescence signal from the glass slide itself. The temptation is to use too much background removal at this stage, because it makes the fluorescence image look better to the eye. If too much background is removed, it is possible that the zero signal level will be set incorrectly by the local background removal algorithm. This has little effect on bright spots but can have a major effect on the integrated fluorescence value reported for dim spots. A good rule of thumb is to set the detector offset so that no (or very few) pixels in the regions between probe spots read zero. MACROview allows the operator to click on any pixel in the image using the mouse, reporting the pixel position and fluorescence intensity at that position. The operator can quickly check to see how much global background subtraction has been used during image acquisition and can set the local background subtraction accordingly.

Finally, the integrated intensity data from each spot on the microarray is stored in a spreadsheet, which can be displayed or compared with data from previous experiments.

3. Conclusion

Confocal imaging provides the resolution, sensitivity, background rejection, and dynamic range required for fluorescence imaging of genetic microarrays. In this chapter, we have reviewed the principles of confocal microscopy and described the confocal MACROscope on which the DNAscope is based. We have shown an example image of a yeast array, and described the steps for analyzing the image.

4. Notes

1. GeneChip is a registered trademark of Affymetrix.
2. MACROscope is a registered trademark of Biomedical Photometrics.
3. MACROview, DNAscope, and *GeneFocus* are trademarks of Biomedical Photometrics; *GeneFocus* is a division of Biomedical Photometrics.
4. The entrance pupil of a laser scan lens is outside the lens body, which allows scanners to be physically positioned at the entrance pupil position. In a microscope objective, the entrance pupil is inside the lens body, so additional optics must be used to translate the scanning beam from the last scan mirror to the entrance pupil. The laser scan lens used is both telecentric and f*theta (f*θ). In a telecentric lens, the converging cone of rays that focuses on the specimen always remains perpendicular to the focal plane, and the focal plane is flat. This means that the fluorescence sensitivity is the same at the edge of the field of view as it is at the center. In an f*θ lens, the distance of the focal spot from the center of the field of view is proportional to the focal length (f) times the scan angle (θ). Thus, if the scanning mirrors are moved at constant angular velocity during the scan, and data are recorded at a constant rate, the recorded pixels will be equally spaced. In a scanning laser microscope, in which the microscope objectives are not f*θ lenses, the distance from the center of the field of view to the focal spot is proportional to f*$\tan\theta$. The result is a different pixel spacing at the edge of the field of view from that at the center. This is not usually a big problem in a microscope, in which scan angles are small ($\tan\theta$ is approximately equal to θ for small values of θ), but is very important in the DNAscope, in which scan angles are large.

Acknowledgments

We wish to thank our colleagues at the Toronto Microarray Consortium for providing the yeast array for imaging, which was printed using a Robotic Spotter designed by Engineering Services Inc. of Toronto.

References

1. Pawley, J. B. (ed.) (1995) *Handbook of Biological Confocal Microscopy*, Plenum, New York.
2. Dixon, A. E., Damaskinos, S., Ribes, A., and Beesley, K. M. (1995) A new confocal scanning beam laser macroscope using a telecentric, f-theta laser scan lens. *J. Microsc.* **178,** 261–266.
3. Dixon, A. E. and Damaskinos, S. (1998) Apparatus and method for scanning laser imaging of macroscopic samples. *US Patent # 5,760,951.*

17

Business Aspects of Biochip Technologies

Kenneth E. Rubenstein

1. Background, Definitions, and Context

In the context of this chapter, biochips are defined as microscale bioanalytical devices that incorporate microfluidic circuitry, highly parallel functionality, or both. Microfluidic circuitry has yielded the lab-on-a-chip concept in which functions such as sample processing, reagent combining, component separation, and detection all occur in sequence on microscale integrated devices. Highly parallel functionality manifests in DNA microarrays, which often contain many thousands of addressable DNA fragments, each capable of hybridizing with a complementary target.

Most biochips are made in processes that borrow one or more steps from semiconductor fabrication technologies. Microfluidic circuitry evolved from work in research groups led by Andreas Manz at Ciba-Geigy *(1)*, Jed Harrison at the University of Alberta *(2)*, and J. Michael Ramsey at the Oak Ridge National Laboratory *(3)* (references represent early work). Mature devices in this category have reservoirs for reagents and waste, electrophoresis columns, cross-channels for reagent introduction, and electroosmotic valving and pumping functions.

DNA microarrays evolved primarily from work done at Affymax, a drug discovery company from which Affymetrix was later spun off, by Fodor and coworkers *(4)*. They combined photolithography with photoactivated oligonucleotide synthesis to generate arrays containing thousands (now hundreds of thousands) of short DNA strands with defined sequence and location.

Microfluidics-based devices have been used to study enzyme reactions, ligand-receptor interactions, and cell-based processes. They find utility in a variety of circumstances, notably high-throughput screening of drug candidates and DNA sequencing. DNA microarrays hybridize with target nucleic acid

From: *Methods in Molecular Biology, vol. 170: DNA Arrays: Methods and Protocols*
Edited by: J. B. Rampal © Humana Press Inc., Totowa, NJ

molecules, usually polymerase chain reaction (PCR) amplicons, in order to study gene expression patterns in cells or to detect genetic polymorphisms. DNA microarrays require specimens that have undergone extensive sample preprocessing, including removal and cleanup of DNA from cells, target amplification, and labeling. The complexity of sample processing has stimulated efforts to integrate microfluidic-based sample-processing stations with the microarrays.

2. Commercial Origins, Activities, and Segmentation

The business of biochips originated, and largely remains, with venture capital-funded startup companies, some of which have gone public, received significant equity participation from large corporations, or both. The biochip business divides neatly into the aforementioned microfluidics and microarray technology segments. Early players on the microfluidics side include Caliper Technologies, ACLARA BioSystems (originally Soane BioSystems), Micronics, Orchid Biocomputer, and Cepheid. Caliper and ACLARA focus on high-throughput screening and DNA sequencing applications, whereas Micronics and Cepheid direct their efforts primarily at in vitro diagnostics. Orchid started with a massively parallel combinatorial synthesis program (funded jointly with SmithKline Beecham), but has more recently emphasized high-throughput screening and genotyping applications.

Significant participants in the microarray technology segment include Affymetrix, Incyte, Hyseq, Molecular Dynamics (now part of Amersham Pharmacia Biotech), Nanogen, Protogene, and Genometrix. Affymetrix and Protogene produce synthesized oligonucleotide arrays whereas the others work with spotted arrays, a variety capable of incorporating larger DNA fragments. All these companies now emphasize genotyping applications. Affymetrix, whose unique patented array-making process permits the incorporation of hundreds of thousands of oligonucleotide spots (or features as Affymetrix calls them), is the only one of these companies able to make sufficiently dense arrays for single nucleotide polymorphism (SNP) discovery. Affymetrix and other companies, including Nanogen, Hyseq, Perkin-Elmer, and Protogene, make smaller arrays for scoring known polymorphisms.

Affymetrix, Incyte, Hyseq, and Nanogen, all from the microarray category, are public companies. Molecular Dynamics was a public company prior to its acquisition by Amersham Pharmacia Biotech. Incyte and Hyseq, both of which feature microarray technologies, identify themselves primarily as genomics companies. Incyte utilizes arrays for its internal work and offers array-based products and services through a subsidiary. Hyseq, which also utilizes arrays for internal purposes, has its microarray-based HyChip system under codevelopment with Perkin-Elmer for genotyping applications. Affymetrix, which sells arrays to the research market directly and to corporate customers on a contract basis, also sells instrumentation and software to make a complete analytical system.

Nanogen's applications portfolio is highly diverse and includes a major partnership with Becton Dickinson for infectious disease diagnosis and a genotyping program directed at both gene discovery and forensic applications. Each oligonucleotide or DNA fragment in a Nanogen array is built on an electrode that is used to manipulate local charge to assist hybridization of targets and dehybridization of mismatched DNA. This arrangement, although permitting significant protocol advantage, also adds complexity in the design of dense arrays. The company is currently working to break the 1000-feature barrier. Nanogen's first product, an instrument and kit for producing custom arrays, is directed at the general research market.

Caliper, in partnership with Agilent, is approaching the pharmaceutical research market with a "personal" research system comprising a relatively inexpensive instrument plus a variety of disposable microchannel cartridges for enzyme assays, ligand-receptor assays, DNA fragment separation, and cell-based assays. Whereas Caliper and ACLARA are clearly producing lab-on-a-chip systems (another name for Manz's *[1]* micrototal analytical system), Cepheid has opted to use microchannel technology and related semiconductor technologies to produce modular instrument-plus-consumable systems built around a microscale PCR thermocycler. The company emphasizes infectious disease diagnosis.

3. Public Sector Support of the Biochip Business

The majority of biochip companies, from both the microchannel and microarray segments, are working under multimillion dollar contracts or grants from the US Department of Defense's Defense Advances Research Projects Agency (DARPA) on portable systems to detect biological warfare agents in the field. ACLARA, e.g., is developing on-chip reagent storage integrated with sample preparation. Cepheid is a key member of a consortium that will receive $5 million over 3 yr to develop a handheld pathogen detection system. Nanogen is working under a 2-yr $2.8 million contract to develop an integrated miniature pathogen detection system, and Orchid has a $12 million contract to develop an integrated microdevice for genetic analysis.

The common element in all these DARPA arrangements is the integration of sample preparation and analysis. Nanogen and Orchid, arguably, are leaders in this realm. Nanogen is working toward using its "bioelectronic" arrays to facilitate cell lysis, separation of nucleic acids from proteins, and other sample-processing steps *(5)*. Orchid has designed a sample-processing front end for its microfluidic-fed microwell systems. The proposed system, which has been partly implemented, traps microbes from the air with antibodies, lyses them, captures the nucleic acids on magnetic particles, amplifies the targets, and transports them via a microfluidic network to a microwell-array for analysis.

DARPA is not the only government agency offering significant support to the development of biochip technologies. The National Institute of Standards and Technology (NIST), the former National Bureau of Standards, offers grants for the development of miniature DNA diagnostic systems through its Advanced Technology Program (ATP) system. In late 1998, ACLARA was the recipient of such a grant, providing $3.6 million for an integrated sample preparation system. Nanogen has received $4 million for an integrated DNA analytical system, and a consortium involving Perkin-Elmer and 3M received $21 million for a similar task. NIST chose Caliper for a $2 million grant aimed at developing a centralized laboratory DNA diagnostic system. Similar grants have come from the Department of Energy (the US home of the Human Genome Project) and the National Institutes of Health.

4. The Role of Large Corporations

Commercial development programs based on biochip technologies have gained sufficient credibility to attract significant collaborative involvement from large corporations. Early buy-ins, before 1998, included BioMerieux's collaboration with Affymetrix on microarrays for the identification of panels of pathogenic bacteria coupled with antibiotic susceptibility testing, Hoffmann-La Roche's suppport for Caliper's high-throughput screening system, SmithKline Beecham's collaboration with Orchid on a combinatorial library microsynthetic system, Becton Dickinson's joint development program with Nanogen on infectious disease diagnosis, and Perkin-Elmer's collaboration with Hyseq on the HyChip system.

During 1998, the pace of deal-making increased in both frequency and scope. Notable for its scope is Hewlett-Packard's collaboration with Caliper, which involves a $20 million joint investment during the first year, to develop a "personal computer" desktop analyzer for the pharmaceutical market. Smaller in scope (precise figures are not available), but possibly greater in ultimate significance, is Perkin-Elmer's joint program with ACLARA for a microchannel-based high-throughput screening system.

The newly emerging generation of DNA sequencers, intended for industrial strength automation, is based on capillary electrophoresis. Systems based on microchannel electrophoresis with at least partially integrated sample processing could become the next generation of high-speed "personal" sequencers. Such systems could be dedicated to a single function or to multiple functions, as is the case for the proposed Hewlett-Packard-Caliper device, which counts DNA sequencing among its targeted applications. A series of chips, each dedicated to a particular application, can be supplied to users, much as Nintendo supplies multiple game cartridges designed to work solely with its hardware.

Perkin-Elmer and Motorola, arguably, are the most highly committed among large corporations to biochip technologies. Perkin-Elmer, probably the largest company in the world dedicated primarily to bioanalytical systems, has parlayed a leadership role in DNA sequencing and a license from Roche for research applications of PCR technology into a strong base from which to acquire and pursue highly miniaturized technologies for DNA sequencing, high-throughput screening of drug candidates, and genotyping. In addition to the aforementioned involvement with ACLARA, Perkin-Elmer is developing several relevant technologies including Hyseq's microarray-based genotyping system, a ligation-based solution genotyping system that uses ZipCode™ microarrays for addressing components, and Taqman™ genotyping technology in highly miniaturized microtiter plates. The addition of 80% ownership in the nascent Celera Genomics venture holds the potential to make Perkin-Elmer into a formidable contender for a key position in the emerging drug discovery paradigm.

Motorola's significant commitment to biochip technology makes it the only company with significant semiconductor involvement to make the apparently obvious leap from computer chips to biochips. A joint venture involving Motorola and Packard Instruments has licensed technology from the Argonne National Laboratory for gel-based microarrays, which have the apparent advantage that hybridization reactions occur essentially in solution rather than at a solution-solid interface. Motorola manufactures chips, and Packard manufactures and distributes ancillary instrumentation. Through its Biochip Systems group, which reportedly employed about 60 staffers at the end of 1998, Motorola has entered into a collaboration with Orchid aimed at enhancing the functionality and manufacturability of the latter's chips and to further develop microfluidic chip technology. Motorola will provide engineering and manufacturing expertise while gaining access to Orchid intellectual property relating to chips, portable diagnostic systems, and certain industrial applications. Motorola will also assist Orchid in the construction of a research and development chip fabrication facility located near Orchid's Princeton, New Jersey, facility. Motorola has also licensed microarray-related technology from and made an equity investment in Genometrix, a pioneer in spotted DNA microarray technology.

Motorola's plans to build a comprehensive internal business are perhaps best illustrated by an employment advertisement in the March 26, 1999, issue of the journal *Science*. Motorola Biochip Systems is "enabling new paradigms in human health care, agriculture, and environmental management." The company is "applying the latest microfabrication, electronics, and information technologies to the cutting edge of biotechnology." Disciplines to be staffed include genomics, molecular biology, bioinformatics, business development, licensing and intellectual property, nucleic acid chemistry, microfluidics, industrial design, automation engineering, and production engineering. The avertisement demonstrates clearly that Motorola is the only large company to form an integrated biochip subsidiary. The emphasis on production of biochips implies that some of the small biochip chip may come to Motorola, as Orchid has already done, for expertise and participation in chip manufacture. Signs indicate that Motorola is aiming to become the "Intel-of-the-biochip-business."

In September 2000, Agilent Technologies announced its entry into the DNA microarray market.

5. Drug Discovery: A Key Factor in the Growth of the Biochip Business

During the mid-1990s, biochip companies tended to organize as venture capital–backed startups with business plans directed heavily toward diagnostic applications. As the realization grew that the market for miniaturized genetic analysis equipment was not yet ready for prime time, these companies redirected their attention toward the revolutionary changes occurring in the pharmaceutical industry. These changes reflect the convergence of several forces including a plethora of patent terminations, the rise of combinatorial chemistry and high-throughput screening technologies, and fallout from the Human Genome Project.

An unusually large number of drugs are on the verge of going off-patent in the early years of the coming decade, a fact that places great pressure on pharmaceutical companies to increase the supply of new drugs emerging from research and development pipelines. A concomitant need to lower the costs for drug development, currently somewhere between $300 and $500 million per drug, is thought to be addressable by eliminating nonviable candidate compounds at earlier than current stages of the development process. Such increases in development and hit rates require more pharmacological targets, more candidate compounds, and better systems to identify problem drug candidates in preclinical development.

Combinatorial chemistry produces large compound libraries and high-throughput screening determines their activity with appropriate pharmacological targets. Concerning target variety, the Human Genome Project, together with its associated private sector correlates, has resulted in the discovery of thousands of new genes. Downstream activities have, in recent years, turned toward functional genomics, in which the biological significance of the newly discovered genes is delineated.

Biochips, in the form of DNA microarrays, are already playing a central role in comparative gene expression studies aimed at elucidating key genetic pathways that are activated in various diseases and in response to various therapeutic modalities. Establishing the identity of genes that are upregulated or downregulated when tissues from normal individuals are compared with those with particular diseases is providing valuable clues to the mechanisms of diseases with genetic components associated with their pathology. Similarly, once gene expression patterns relating to particular disease phenotypes have been established, they can be used as multiparametric measures of efficacy for new drugs or therapeutic modalities.

DNA microarrays are also playing a role, which may become central, in genotyping studies for the discovery and scoring of SNPs. Such studies are expected to identify genes connected to the susceptibility for various genetically complex diseases such as diabetes, schizophrenia, and high blood pressure. The discovery of relevant genes can identify new pharmacological targets with levels of accuracy and speed never before attainable.

Perhaps the most exciting application of microarray technology lies in scoring SNPs and other genetic polymorphisms in individuals. The emerging field of pharmacogenomics promises to produce drugs tailored to groups of individuals who share common polymorphisms. Consequently, a dynamic new area of commerce and medicine, combining diagnostics with therapeutics, promises to emerge in the next 5 yr. An individual with schizophrenia, e.g., may be tested for one or more particular SNPs that point to the prescription of a particular drug that is both safe and effective for people in that genetic category.

Because the genetically complex diseases include highly prevalent ones, the market potential for such testing is impressive. For example, a conceivable estimate for the worldwide market might include 10 diseases, for each of which 5 million individuals are tested per year at a cost of $20 per microarray per individual. This estimate leads to $1 billion per year in revenues for microarray manufacturers. The possibility of integrating sample processing with microarray analysis for some subset of these tests adds a further significant multiple. Of course, genotypes do not change, so each individual would be tested only once in a lifetime for any given polymorphism. On the other hand, gene expression studies of the type mentioned earlier might be performed repeatedly on individuals in order to monitor therapy with particular drugs.

6. Intellectual Property Conflicts

The technological power and commercial significance of DNA microarrays are clearly evident, but the attendant competitive scenario is complex and subject to considerable uncertainty. Patents contribute heavily to this complexity.

The scientific founders of Hyseq generated patents, based on work done originally in Yugoslavia and later at the Argonne National Laboratories, on the concept of sequencing by hybridization. These patents were licensed to Hyseq and form the basis of a patent infringement lawsuit filed against Affymetrix in 1998. Affymetrix has since countersued Hyseq for infringements of its intellectual property. Although the possibility of serious repercussions for the balance of competitive power in the microarray field is probably remote, it is well to remember that similar cases have made significant impacts in other markets. Participants in the immunodiagnostics market of the early 1980s may remember the case of Hybritech vs Monoclonal Antibodies based on the former's patent covering sandwich immunoassays using two monoclonal antibodies. Hybritech lost the infringement suit, but prevailed on appeal and for a time dominated a market in which Monoclonal Antibodies quickly became an insignificant participant.

Further intellectual property complexity derives from a patent granted to Affymetrix covering two-color measurements for spotted microarrays. The patent application, which was filed based on work deriving from a partnership between Incyte and Affymetrix, covers basic technology at the heart of Incyte's microarray system. Affymetrix has filed a patent infringement lawsuit against Incyte and granted licenses for spotted array technology to several companies, including Molecular Dynamics. Controlling access to major sectors of both spotted and synthesized array technologies would provide Affymetrix considerable competitive advantages. Yet, the Hyseq challenge, which covers the very basis of microarray technologies, must be considered a serious one.

Although no patent litigation has yet emerged on the microchannel side of the biochip market, it can certainly be expected once products enter the marketplace and revenues grow to appreciable levels. Several companies share common microfluidics technologies such as microchannel electrophoresis and electrokinetically emulated valves and pumps. Caliper Technologies has licensed intellectual property deriving from the work of J. Michael Ramsey at Oak Ridge National Laboratory, and ACLARA has its own patents in the area. The two companies are moving in highly parallel directions, and some form of legal collision must be considered likely.

7. Future Trends

Legalities aside, the future of microchannel-based systems rests on two major attributes: speed and functional integration. Speed may prove valuable in two application areas: DNA sequencing and high-throughput screening. The life sciences market would appear to be highly receptive to an ultrahigh-speed sequencing system based on disposable microchannel cartridges, each containing multiple electrophoresis lanes.

The case for speed in the high-throughput screening application is less clear. The field began with automated robotic systems based on 96-well microtiter plates and volumes in the 0.5-mL range. The pharmaceutical market has demanded smaller volumes and higher throughput, and suppliers responded with plates containing 384 wells and considerably reduced volumes. Microplate technology is currently moving to 1536 wells and volumes in the microliter range. These volumes are perhaps small enough to satisfy industry demands for minimizing the consumption of valuable materials. The microchannel systems, using Caliper's high-throughput screening system, e.g., operate on picoliter volumes, run reactions very quickly, avoid external pipetting of reagents other than the specimen, and are primarily sequential rather than parallel in operation. Although head-to-head comparisons of the two types of systems are not yet available, the microwell systems appear to have become a de facto standard. Not only must the microchannel systems be justified on the basis of relative attributes and performance, but they must displace an increasingly entrenched technology base, which is supported by multiple manufacturers and already active in the marketplace.

Market demand for functional integration, the other potential attribute for microchannel systems, must also be considered the subject of considerable uncertainty. The lab-on-a-chip concept implies that multiple processes, formerly requiring separate manual steps or separate instrument modules at the macroscale, can be performed with few, or no, moving parts at the microscale in a single disposable cartridge. Caliper and Agilent are betting that the concept of a personal workstation capable of operation with a variety of cartridge designs, each representing a different analytical process, will capture significant market attention, particularly from the pharmaceutical industry.

Personalization or decentralization certainly worked in the case of personal computers, which co-opted many functions previously relegated to mainframes or minicomputers. The decentralization concept has been less than overwhelmingly successful when applied to point-of-care medical diagnostics. The idea was to bring diagnostic testing closer to the patient in both space and time, just as the concept in microchannel research instrumentation involves bringing analytical capability closer to the researcher in space and time.

During the past two decades, many venture-backed companies attempted to apply the principles of miniaturization and functional integration to this diagnostic market. From both the investment and market perspectives, these efforts were quite disappointing. The systems took considerably longer to develop than originally predicted and often performed somewhat less well than intended. Perhaps the most significant issue is that decentralization of testing also meant decentralization of selling. Many customers had to sell on the idea of approaching their work from a different perspective. Product attributes, although attractive, were perhaps insufficiently advantageous to convert customers at the rates and in the numbers required. The personal computer provided customers with enormous value. Whether functionally integrated biochips can do the same remains to be seen. As the old marketing adage goes, "We don't sell the steak, we sell the sizzle."

References

1. Manz, A., Graber, N., and Widmer, H. M. (1990) Miniaturized total analysis systems: A novel concept for chemical sensors. *Sensors Actuators* **B1,** 244–248.
2. Harrison, D. J., Fluri, K., Seiler, K., Zhonghui, F., Effenhauser, C. S., and Manz, A. (1993) Micromachining a miniaturized capillary electrophoresis-based chemical analysis system on a chip. *Science* **261,** 895–897.
3. Jacobson, S. C., Hergenroder, R., Koutny, L. B., and Ramsey, J. M. (1994) High speed separations on a microchip. *Analyt. Chem.* **66,** 1114–1118.
4. Fodor, S. P. A., Read, J. L., Pirrung, M. C., Stryer, L., Lu, A. T., and Solas, D. (1991) Light-directed spatially addressable parallel chemical synthesis. *Science* **251,** 767–773.
5. Cheng, J., Sheldon, E. L., Wu, L., Uribe, A., Gerrue, L. O., Carrino, J., Heller, M. J., and O'Connell, J. P. (1998) Preparation and hybridization analysis of DNA/RNA from *E. coli* on microfabricated bioelectronic chips. *Nat. Biotechnol.* **16,** 541–546.

Index